THE VIKING AGE

THE VIKING AGE

A TIME OF MANY FACES

by

CAROLINE AHLSTRÖM ARCINI

with

T. Douglas Price, Bengt Jacobsson, Maria Cinthio, Leena Drenzel,
Bibiana Agustí Farjas and Jonny Karlsson

Illustrated by

Staffan Hyll

Oxford & Philadelphia

First published as a hardback in 2018. Reprinted as a paperback in the United Kingdom in 2022 by
OXBOW BOOKS
The Old Music Hall, 106–108 Cowley Road, Oxford, OX4 1JE

and in the United States by
OXBOW BOOKS
1950 Lawrence Road, Havertown, PA 19083

© Oxbow Books and Caroline Ahlström Arcini 2018

Paperback Edition: ISBN 978-1-78925-804-2
Digital Edition: ISBN 978-1-78570-939-5 (epub)

A CIP record for this book is available from the British Library

Library of Congress Control Number: 2018939501

All rights reserved. No part of this book may be reproduced or transmitted in any form or by any means, electronic or mechanical including photocopying, recording or by any information storage and retrieval system, without permission from the publisher in writing.

For a complete list of Oxbow titles, please contact:

UNITED KINGDOM
Oxbow Books
Telephone (01865) 241249
Email: oxbow@oxbowbooks.com
www.oxbowbooks.com

UNITED STATES OF AMERICA
Oxbow Books
Telephone (610) 853-9131, Fax (610) 853-9146
Email: queries@casemateacademic.com
www.casemateacademic.com/oxbow

Oxbow Books is part of the Casemate Group

Front cover: A male skull with filed teeth from the Viking Age cemetery in Fjälkinge, north-east Skåne in Sweden. Superimposed on the skull is a mask from a runestone in Lund, Lundagårsstenen DR 314. (Photo Staffan Hyll)

Contents

Preface and acknowledgment — vii

1. The bare bones — 1
2. Eight Viking Age burial grounds in south-east Sweden — 5
   - Trinitatis: an early Christian graveyard in Lund — 5
   - Vannhög: a burial place near an old Viking fortress — 9
   - Fjälkinge: a remarkable burial ground on the fertile plain — 13
   - Kopparsvik: a cemetery south of Visby — 17
   - Slite Square: with a view of sailing routes to the east — 20
   - Fröjel: a burial ground beside a Viking Age harbour — 21
   - Birka: a well-known trading place in the realm of the Svear — 28
   - Skämsta: a farm cemetery — 34
   - A wide range of burial practices — 34
   - Everyone was buried — 37
3. Immigrants or locals? — 39
   - A geological signature can be detected in dental enamel — 40
   - Different patterns emerge — 46
   - Someone knew how the deceased wanted to be buried — 49
   - Did everyone come here voluntarily? — 53
4. Health and care for the frail — 55
   - "Tall as palm trees" — 55
   - Toothless or shining white? — 57
   - Joint problems — 58
   - Everyday accidents and battle traumas — 60
   - The dwarf — 64
   - Leprosy: noseless and numb — 67
   - Health in Viking Age society — 70
5. Markers of identity? — 73
   - Filed grooves on the teeth — 74
   - Young, old, short, and tall — 77
   - Buried like other people? — 77
   - Was Gotland the gathering point? — 79
   - A Nordic custom or inspiration from elsewhere? — 80
   - Why file grooves in teeth? — 82
6. Burial grounds designated for particular purposes? — 85
   - The influence of Christianity or division into special areas? — 87
   - Market places and harbours? — 88
7. A time of many faces — 91

Appendix: Strontium values — 95
Notes — 109
References — 113

Preface and acknowledgment

Suffice it to say, the Viking Age represents the most exposed period in the prehistory of Scandinavia, with many academic texts as well as representations in popular culture. Too often, however, the descriptions about this period concentrate on death, violence, raids and fear. However, archaeological and osteological studies have delivered, and continue to deliver, other perspectives regarding Viking Age society. This book is one such example, where the everyday life of people is in focus. It is via the skeletons that we present the life history of children and adults during the Viking Age. As an osteologist working in the field of archaeology I regularly work with human skeletal remains from the Viking Age. The material presented here is based on excavations from the twentieth century. The purpose of the project was to synthesize what the skeletons tell us about the living condition for people during the Viking Age.

Several colleagues have in various ways contributed to this publication. My husband Torbjörn Ahlström, Department of Archaeology and Ancient History, Lund University, suggested already at the beginning of this project that we should use strontium isotopes to map mobility in Viking society. Douglas Price, University of Wisconsin, with his knowledge and experience of this method, has contributed not only to this publication, but also through other studies showing the variation in mobility in Scandinavia during this period. Leena Drenzel, National Historical Museum, Stockholm, has spent several weeks with me in the museum depot in Tumba, where teeth were studied under the magnifying glass. Jonny Karlsson, National Historical Museum, Stockholm, Bibiana Agustí Farjas, Insitu S.C.P. Arqueologia funerària, preventiva i patrimoni cultural Begur/Centelles/Sant Feliu de Guíxols, and Torstein Sjøvold have delivered discoveries of new cases of modified teeth in Sweden. I also thank Johan Calmer and Ingmar Jansson for communicating contacts in Russia and Ukraine in search of possible traces of the phenomenon of modified teeth. Bengt Jacobsson, Riksantikvarieämbetet UV Syd, and Bertil Helgesson, Sydsvensk Arkeologi AB, first introduced me to grave materials from the Viking era. Together with Maria Cinthio, I have had intensive and constructive discussions about Viking burials in a Christian context and especially in Lund. Dan Carlsson, Arendus and Lena Thunmark-Nyhlén have been valuable discussion partners regarding the Gotlandic surveys. Together with Per Frölund, Swedish University of Agricultural Sciences, I have worked with the graves from Skämsta. Rita Larje has provided valuable osteological information about the graves from the cemetery in Kopparsvik, Gotland. With Mattias Toplak, Eberhard Karls Universität, Tübingen, I have had several interesting discussions regarding the phenomenon of prone burials. Gunnar Andersson, National Historical Museum, and Ingrid Gustin, Department of Archaeology, Lund University, have contributed information regarding Birka. I have had interesting discussions regarding strontium with Ola Magnell and Mathilda Kjällquist, The Archaeologists, National Historical Museum, and Helene Wilhelmson, Sydsvensk Arkeologi AB. I have discussed the difference between Slavic ceramics and Baltic ware with Mats Roslund, Department of Archaeology and Ancient History, Lund University, and Torbjörn Brorsson, KKS. I want to thank Göran Possnert, Tandem Laboratory at Uppsala University for radiocarbon dates. I have received valuable comments on the text, especially from my good friend Annica Cardell. The translation into English was made by Alan Crozier. I also want to thank Helene Borna-Ahlkvist at The Archaeologists, National Historical Museum, for support during the whole project. Funding has been provided from the Birgit and Gad Rausing Foundation for Humanistic Research, Ebbe Kock Foundation, Berit Wallenberg Foundation, Gotlandsfonden, Åke Wibergs stiftelse and DBV. I also want to thank Gotlands fornsal in Visby, National Historical Museum, Stockholm, Historical Museum and Kulturen's Museums in Lund and Trelleborg Museum in Trelleborg. Finally I will thank Staffan Hyll who has a fantastic ability to visualize my ideas as comprehensible illustrations.

I dedicate this book in memory of my tutor, colleague and good friend Pia Bennike, who followed and supported this project from the beginning but alas, did not see the final book.

1

The bare bones

Vikings. Say the word and we think of robbery, rape and pillage, assault, battles, kings, chiefs, mercenaries, and colonization. *Vikings*. We think of ships, long-distance travel and connections, runes, buried hoards of silver, and trade in both goods and people. *Vikings*. We think of the meeting between paganism and Christianity. *Vikings*. We think of the Icelandic sagas, of European settlers in Greenland and Vinland and ibn Fadlan's descriptions of the Norse traders in the east, "Rus" as tall as palm trees, with blond hair and tattooed bodies and young female slaves following their masters into death. But there is much more to tell.

Some scholars believe that the Viking Age is not a correct designation because the word Viking mainly denotes a "pirate" and describes only that part of the population who set off on plundering expeditions.[1] Nor was there any homogeneous Viking Age culture or a Viking Age people, since Scandinavia at this time consisted of small geographical areas with different names and traditions (Fig. 1), not countries with borders like the present-day Norway, Sweden, and Denmark.[2] On the other hand, the Old Norse mythology and language were the same.[3]

Although the word Viking is mainly associated with the more violent activities of this period, precious metals in different shapes and labour in the form of slaves were brought home and thus also benefited the more sedentary part of the Norse population. In other words, the results of the actions performed by the Vikings were deeply integrated in the society of the time. The traces of the period we today call the Viking Age, AD *c.* 750–1050, represent both those who travelled on raids and military expeditions, and those who peacefully pursued trade and made their living from what the farm yielded. Representatives of the different groups were, in all probability, intimately associated with one other. So, is it really possible to distinguish between those who are called Vikings and the rest of the population?

This book deals with both those who lived their entire lives in Scandinavia and with those who set off *i víking*, "on plundering expeditions". The latter usually came back, perhaps switched to another livelihood, and were buried at home. The findings presented here are based on the skeletons of those who originally came from Scandinavia. But as we shall see, there were also foreigners in these places of burial. Clues to a knowledge of this period come from archaeological excavations of burial grounds. Here we *find* the skeletons of people of all ages from different social classes. Studies of the skeletons not only give us insight into how people expressed their identity but also enable us to detect the situation of the most vulnerable individuals in society, namely, children and the elderly. We get a glimpse of health conditions in those days, and how social networks functioned, and of how people viewed and treated those who differed in various ways.

Previous scholarly studies of Viking Age populations in Scandinavia, based on analyses of skeletal material, have been published by a number of researchers.[4] These studies deal with matters of health and migration, and the fact that a proportion of the population consisted of slaves. Researchers have also examined the dental health of the Viking Age population and found a high degree of tooth loss and heavy wear, along with a widespread prevalence of caries, but there are also examples of attempts at dental hygiene reflected in the use of toothpicks.[5] A study of cemeteries in the Mälaren valley suggests, for instance, that the health status of the people buried in the cemeteries adjacent to the well-known trading site of Birka seems to have presented more infections and disturbances in nutritional intake than, say, those buried

Fig 1 Map of Sweden with known names of historical areas. (after Fredrik Svanberg 2016 and sources cited therein)

in the early Christian graveyards in Sigtuna.[6] Dietary analyses of the early Christian burial places in Sigtuna also show that the women consumed more vegetables than men.[7] That people did not only travel *from* but also *to* Scandinavia is seen, for example, in an aDNA analysis of the skeletons in the Viking Age Oseberg ship in Norway. The young woman buried in the ship may even have come from as far away as the region around the Black Sea.[8] Even more examples of mobility – perhaps not always voluntary – are indicated by the study of a dozen or so graves at Flakstad in Norway. Isotope and aDNA analyses led to the conclusion that some of the individuals were slaves.[9] Another recent study of conditions in the Iron Age concerned the population of the island of Öland in the Baltic Sea. Based on strontium analyses, this study demonstrates that immigration of people to Öland was greater in the Viking Age than in earlier periods of the Iron Age.[10] It was also observed that the origin of the newcomers varied, from nearby to far away.[11] Large-scale strontium analyses of Viking Age grave material from Denmark has also shown that those who came from outside tended to be buried with more grave goods than the local people,[12] and that a large number of the men in the military forces of Harald Bluetooth, who reigned in the 10th century, were recruited outside Denmark.[13.] Our study, however, is the first synthesis of its kind examining Viking Age populations from several different regions in present-day southern Sweden.

During the Viking Age there were two different ways of handling the bodies of the deceased: they were either burnt on a pyre (cremation) or buried unburnt (skeletal graves). To some extent these burial practices occurred in parallel depending on where in Scandinavia we look, but a certain chronological shift can be observed, as cremation graves dominated in the earlier phase of the Viking Period. In several of the geographical areas investigated here there was a switch to skeletal burials early in the Viking Age. The graves used in this study are exclusively skeletal. The reason for this is that one of the main purposes was to study health and living conditions in the Viking Age, which is difficult in the case of cremation graves, where diseases cannot be detected in the same way as with a skeleton.

In the chapter *Eight Viking Age burial grounds in south-east Sweden* we present the burial sites and the roughly 1800 graves which form the basis for this study. It will show the geographical distribution of the Viking Age settlement and different burial customs and form the basis for demographic discussions, i.e. age and sex distribution. The majority of all the investigated sites are located near the coast in southern and south-eastern Sweden (Fig. 2). *Immigrants or locals?* is the chapter where we use strontium analysis to study mobility. Strontium analysis of the enamel of individuals is a method that has been frequently used in recent years to discern who is local and who originated from another region. To study and compare the non-locals' burial customs with those of the locals could give us an indication of how integrated the foreigners were in the society and whether they kept some of their customs. This study covering nearly 400 individuals from both rural and early urban Viking societies provides new knowledge about the movement of individuals in a geographical perspective during the Viking Age.

The chapter *Health and care for the frail* deals with the possibility to obtain both knowledge of living conditions in a population and how the relatives and the community cared for the sick and disabled. The material has provided good indications of the health situation during the Viking Age, and we will discuss for example whether the disabled lived long despite their handicap and if they were buried in the same careful way as others. During this period, the

Fig 2 Map of Sweden with the eight burial sites presented.

infectious disease leprosy was introduced to the population, and we have used strontium analysis to investigate whether migration is a likely reason for the spread of the disease. In the Middle Ages people with leprosy were not allowed to live and be buried among the others, but what was the custom during the Viking Age? All this is interesting not least because the Viking Age is often described in brutal terms.

People in all times worldwide have adorned their bodies with temporary or more permanent decorations. The best known and most common permanent decorations are tattoos. However, there is also a permanent decoration that people made by filing their teeth – a custom that we find evidence of among men from Scandinavia during the Viking Age. In the chapter *Markers of identity?* we discuss the possible meaning of this custom and its distribution. The presentation of the different burial grounds demonstrates a very wide variation regarding for example demography and in the chapter *Burial grounds designated for particular purposes?* we discuss different possible causes.

In the final chapter, *A time of many faces*, we summarize how a study like this contributes to our knowledge about people and their living conditions during the Viking Age. As the reader will see, the title is deliberately ambiguous. We already know that people in the Viking Age travelled much and travelled far, which means that they met people from different cultures, who dressed and adorned themselves differently, and differed in the colour of their skin, hair, and eyes. In short, there were encounters between people with different faces. As some of the methods of analysis show, it was not just the case that people travelled from Scandinavia; foreigners came here too. If travellers in the Viking Age came across people with different-looking faces, it could be because they were afflicted with various diseases. In certain cases, unfortunately, it meant that the faces changed beyond recognition. Last but not least, there was the practice adopted by men during the Viking Age, namely, changing the appearance of their front teeth, as a result of which a person who met a Norseman like this would not forget his face.

This publication aims to describe the Viking Age on the basis of new material that has not hitherto received the attention it deserves, namely, the skeletal material. The idea is to synthesize this research and thus further contribute to the accumulation of knowledge about living conditions during the Viking Age. Since this study also covers a relatively large geographical area, we can obtain a good picture of the health status during the period. Behind every skeleton is an individual; some of them lived only a few hours, while others had a long and eventful life. This book will reveal what life was like for them.

Bones and teeth of a person from a bygone age can be preserved for thousands of years. Researchers in different disciplines together contribute to derive new knowledge about the living circumstances of individuals and groups. Varying housing standards, hygiene, food habits, starvation, infections, accidents, and violence, as well as caring and the social safety nets, leave their marks on the skeleton. Assessments of the age and sex of the individuals in this study are based on methods compiled by Buikstra and Ubelaker.[14] For the estimation of the individual's stature, the method of Sjøvold has been used.[15]

2

Eight Viking Age burial grounds in south-east Sweden

The Viking Age was a time of change as regards religion, when pagans became Christians and the old mortuary traditions gave way to a new idiom. People on the Scandinavian peninsula began to abandon the beliefs and customs that are associated with the pre-Christian period. They slowly gave up the practice of cremating their dead on a pyre, and instead buried the body as it was. To some extent the custom of inhumation had already gained a foothold, which did not mean that all the old traditions were abandoned. People were still interred with grave goods of various kinds, personal belongings, and also other objects. Wearing their clothes and, in some cases, ornaments, the bodies were laid either directly in the ground or in a coffin. In special contexts an entire chamber was built of timber,[1] in which the deceased was placed together with various necessities. According to the new Christian faith, with which people at this time were beginning to come into contact, and which was gradually spreading, the dead no longer needed to have food or personal belongings such as jewellery or weapons. With the introduction of Christianity everyone was buried in the same direction, with the head to the west and the feet to the east, ready to meet their maker on the last day. The burial practice was more egalitarian, although those who were better off could choose a fine burial place, for example an area near the most important building of the new faith, the church. In this regard, however, small children were an exception; whatever their social class we can see that in most cases at the oldest churches they were buried around the chancel and along the walls of the church.

The place of burial plays a crucial part in people's everyday lives and history. Knowing where one's kinsfolk are buried is something that affects everyone to a greater or lesser extent and, in historical terms, burial places contribute much of what we know about people's lives and thoughts. The grave is associated with the ceremonies performed by the survivors for generation after generation. Such customs were collectively agreed on and they can, in some measure, be interpreted even several thousand years later. How close the interpretations are to the reality is difficult to say, but there are signs of the occurrence of fundamental values shared by humans no matter in what time they have lived.

Before we look in detail at the changing demographic conditions that burial grounds can reflect, and the causes of this, we should look more closely at the sites that here represent the Viking Age. In this context eight Viking Age burial places from southern and south-eastern Sweden have been analysed and serve as examples (Fig. 2).

Trinitatis: an early Christian graveyard in Lund

On the fertile plain of south-west Skåne lies the university town of Lund, whose history can be traced back to the end of the Viking Age. In the Early Middle Ages the town had 26 churches but the only one of these that survives is the Cathedral. If you could stand on the site of the cathedral looking to the south-west and go back a thousand years in your imagination, you would be able to see the wooden church of Trinitatis and its large cemetery. Several rounds of archaeological excavations have been conducted since the start of the 1940s. Traces of parts of Trinitatis church and over a thousand graves from the Viking Age have been exposed and their contents retrieved. Under the ground there are still the remains of other parts of the church and presumably another thousand or more graves (Fig. 3).

Lund at the end of the Viking Age was a religious and administrative centre and an early hub in the Danish kingdom. The latest research indicates that two kings, a father and his son, set their stamp on the town of Lund. These are Harald Bluetooth (AD 911–987) and his son Sweyn Forkbeard (AD 960–1014). Harald, unlike his father Gorm

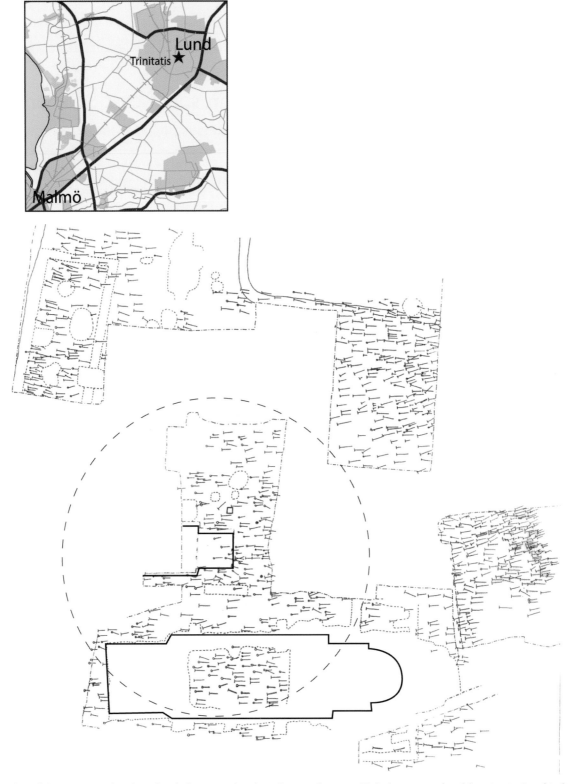

Fig 3 Plan of the Trinitatis churchyard with the stone church and its predecessor. With the geographical location in Lund indicated.

the Old, did not just accept Christianity but also declared, on the runestone from Jelling, that he Christianized the Danes.[2] During Harald's time the kingdom of Denmark was a loosely joined area where the king ruled but the informal power lay with a number of chieftains. Harald was the factor unifying different regions and groups of magnates in what researchers refer to as an early state. His primary role was to serve as commander in chief and preserve peace in the region, besides being the protector of the church. Through his retinue of warriors, the *hird*, he led both trading and plundering, which generated a surplus that was redistributed to subjects at different levels. Through alliances and marriages Harald had established contacts with the realms of the Wends on the southern coast of the Baltic Sea.

The carefully made log coffins found by early excavations north of Lund Cathedral are evidence that Harald may have built an early church and thus begun the conversion of this part of Skåne. The wood in one of the log coffins has been dated by dendrochronology to AD 979–980.[3] This does not mean that people in other parts of Skåne had not already come into contact with the Christian religion and even adopted it. At the end of Harald Bluetooth's reign his son Sweyn Forkbeard took over, and during his lifetime he became king not only of Denmark but also of Norway and, for a short time, England too. It is stated that Sweyn Forkbeard built a church in Skåne, a church which may be the one called Trinitatis. No physical remains of the church survive, just the pits left by the posts that were the frame of the building. The dating of the church and of the graves is therefore based on wood from coffins in the surrounding churchyard, which has been dendrochronologically dated to the start of the AD 990s.

The churchyard of Trinitatis came into existence during the last century of what we call the Viking Age. Parts of the graveyard were used until the Reformation in 1536. The part involved in this study covers only the period 990–1050/60. The wooden Trinitatis church was replaced after a time by a stone church, construction of which is thought to have started around AD 1020/30, or 1050. The question of who used the Trinitatis graveyard is debated. Some scholars think that it was used by Christians from different parts of Skåne,[4] while others are of the opinion that the mode of burial in its oldest phase indicates that those who were buried there largely originated from outside what was then Denmark.[5] Lund is said to have been established as a Christian centre, with a royal mint. In connection with this, people from the south coast of the Baltic, from Germany, and from England all settled there.[6] A closer study of mortuary practice[7] suggests that the cemetery was socially and culturally divided. According to the provincial laws, the area closest to the church was reserved for people of the church and those high up in the social hierarchy while the peripheral areas of the cemetery were used for those of lower social standing.[8] According to researchers, the burials also indicate that different parts of the cemetery were used for people of different cultural origin.[9] Those buried at the outer edge of the cemetery were in coffins constructed in a different manner and in discarded and patched-up coffins made from old troughs (Fig. 4).[10] Moreover, these makeshift coffins were not always long enough for the individual who was to be buried, showing that people took whatever they had.

On the periphery of the graveyard it also happened that people were buried in coffins made of boat planks which had been sealed with sphagnum, other coffins had ribbed bottoms and, in one case, the coffin was made of wattle. One woman was buried in a large withy basket. In several instances the coffins were lined with grass or straw. A majority of these bodies had been given hazel rods in the grave. This paints a picture of very diversified burial customs with indications of influence from Slavic areas, that is to say, from the countries around the southern Baltic. The presence of people in Lund from Slavic areas is also indicated by large amounts of finds of the pottery known as Baltic ware (Fig. 5), made in Lund but by people of Slavic origin.[11] In the area closest to the church the mortuary practice was different. Here many bodies were buried without a coffin, and some lying on a bed of charcoal or lime (Fig. 4). Placing charcoal in the grave was a custom that may have been intended to distinguish the grave for eternity. The tradition is known from prestigious English graves.[12]

The majority of the bodies were laid on their back with straight legs, but occasional individuals were buried with their legs slightly bent. One person in the area nearest to the church was interred in an extremely strange position. The skeleton was found in a rectangular coffin that was half the normal size. The deceased was an elderly woman who was folded with her legs bent backwards from the hips and stretched under the body so that her feet lay above her head. Anatomy makes it impossible to fold a person at the hips unless the body is completely putrefied, and in that case the bones ought to have been in such a state that they would have been separated from each other, but this did not happen. Nor can it be a matter of a skeletonized individual who was buried with such precision. The explanation suggested is that this is a person who was travelling far from the place where she wished to be buried and therefore, after some decay had set in, she was packed and salted in the little wooden crate in which she was buried.[13] The body must have been partially decomposed. The transport of salted corpses is known from written sources.[14]

All the people interred in the Trinitatis cemetery were Christians and were therefore laid in an east–west direction with the head to the west. As Christians, people no longer buried their relatives with grave goods as they had done in pagan times. In some cases, however, there were departures from Christian practice. In one grave there was a little hoard of silver beside the hip of a skeleton. It contained coins

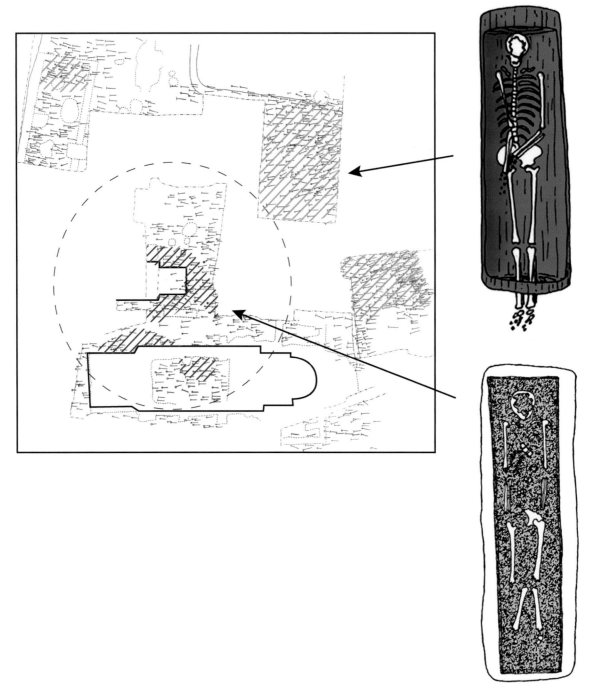

Fig 4 Plan of excavated parts of Trinitatis churchyard. Individuals buried near the church (within the circle) and individuals buried in the periphery.

minted in England, Germany, and Lund. The dating of the roughly 30 silver coins indicates that the body was buried in AD 997 or shortly afterwards. Identification and dating were performed by Jørgen Steen Jensen.[15] Objects of an even more personal nature could accompany their owner, such as a finger ring of gold, an amber pendant, a bead necklace, or a comb. One person was also given a knife, another was given a yardstick, and a third had something that is interpreted as a malt stirrer for beer brewing. Those buried without a coffin may, in some cases, have been interred in their clothes, as evidenced by finds of shoes.[16]

During the period when these burials took place, around AD 990–1050/60, several thousand inhabitants of the new town of Lund were given their last resting place in this cemetery. Several graves were wholly intact while others were damaged by later burials. During the 20th century it happened several times that archaeological excavations took place in parts of the early cemetery. In this connection, a total of more than 1100 skeletons have been unearthed. The age distribution shows that the proportion of children, especially infants, is small (Fig. 6). This applies in particular in comparison to a totally excavated cemetery close by (Kattesundskyrkan K3, AD 1050–1100).[17] As for the sex distribution, this is somewhat skewed, with more men than women (Fig. 7). The reason for the small number of children is something we shall return to in Chapter 3. It is important to bear in mind, however, that only 40% of the cemetery has been completely excavated, but the wide distribution of the excavated parts can probably be regarded as representative of the rest of the cemetery. The excavations were performed, and the material has been analysed on various occasions by archaeologists working for the Kulturen museum in Lund.[18] The osteological analysis of the skeletons was undertaken by an osteologist at Kulturen.[19]

Vannhög: a burial place near an old Viking fortress

On the south coast of Skåne, in the shallow bay near the southernmost tip of Sweden, is the town of Trelleborg. Here the brackish water of the Baltic Sea meets the saltier water of the Öresund. Trelleborg today is the second biggest port in Sweden, with ferries running to Germany and Poland. In the north-west part of the town there is a school named

Fig 5 A vessel of Baltic ware found in Lund. (Photo Viveca Ohlsson, Kulturen, Lund)

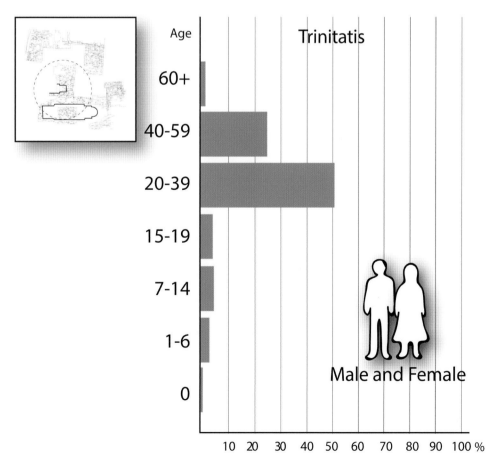

Fig 6 The age distribution at Trinitatis, note that the proportion of children, especially infants, is very small.

Vannhög and just south of it a number of farms in the Viking Age marked out a piece of land which was traditionally used as a burial place for their dead kinsfolk. Thanks to the archaeologist Bengt Jacobsson and his team, the present-day town of Trelleborg now has a rare reconstruction from Old Norse times, a ringfort from the reign of Harald Bluetooth (Fig. 8).[20] For a long time it was only the ringforts in present-day Denmark – such as Aggersborg, Fyrkat in Jutland, Nonnebakken on Fyn and Trelleborg on Sjælland – that were known. The fort in Trelleborg was built in two phases and it is phase two that coincides with the reign of Harald Bluetooth, AD 958–986[21] and the construction of the Danish forts. The reason why Harald Bluetooth built the forts is debated, and before the Danish examples could be dated to show that they came into existence as early as the 980s, the theory was propounded that they were connected to the revolt of Sweyn Forkbeard. They are said to have been training camps and winter quarters for the Viking army. According to other scholars, they were part of the defence of the kingdom and served as a refuge for the local people in the event of attack,[22] whereas Christiansen thought that they existed to control and unite the kingdom.[23] Yet another theory was that they were built for the king's administration and production of commodities for his own needs.[24] The fort in Trelleborg, however, has an earlier phase, phase one, which may have been constructed in the first half of the 10th century. Jacobsson is of the opinion that the original fort might have been purely defensive in character, to protect the local population in times of unrest.[25] South of the fort in Trelleborg, excavations have demonstrated the existence of a shoreline settlement going back a long time, which began to expand in the Vendel Period (AD 550–800). The results, however, suggest that the place was abandoned by the 11th century. Coastal settlements were no longer considered safe, so people moved further inland.[26]

But where did people bury their dead? Close to the fort only three skeletons have been found, in the Katten block

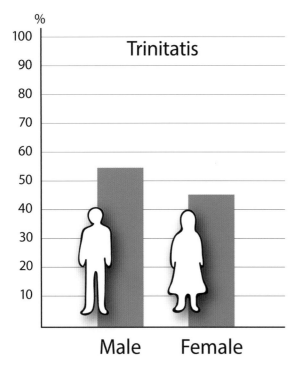

Fig 7 The sex distribution at Trinitatis. It is skewed, with more men than women.

Fig 8 A reconstruction of a ringfort from the Viking period, during the reign of Harald Bluetooth. (Photo Mats Larsssson, Trelleborg)

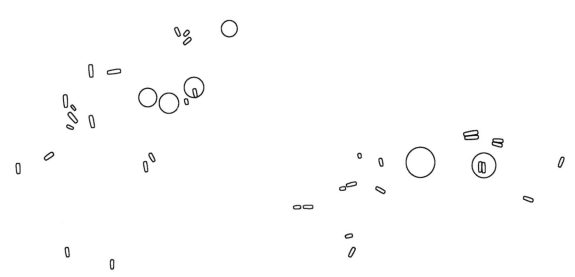

Fig 9 Plan of excavated area of the cemetery and the graves of Vannhög, with the geographical location.

south of the fort. The bodies were those of two men and a small child. According to the register of ancient monuments, there are several gravefields of varying size around the town. One of the larger excavated burial grounds is Vannhög, about 1 km north-west of the fort (Fig. 9). The site has been excavated several times.[27] It has not seen total excavation, which means that we do not know where the boundaries of the cemetery ran, and therefore cannot say how many people were originally buried here. About 40 graves have been excavated and roughly the same number have been damaged by quarrying of gravel at the site.[28] Scattered skeletal parts have been retrieved from these damaged graves. It is difficult to say who the burial ground was used by, but its size indicates that it cannot have been just one farm, more likely several farms together marking and tending an area for their dead. The total number of skeletons examined is 47. The osteological analysis of the skeletal material for this study was performed by Arcini. The age distribution of the bones does not comprise any infants, but the number of children aged 1–6 and 7–14 is what can be expected (Fig. 10). As for the sex distribution, this was not balanced, with twice as many men as women (Fig. 11).

Based on the finds and four radiocarbon dates, the burial ground was judged to have functioned roughly from the end of the 7th century to the middle of the 10th century. Preservation conditions for wood were poorer here than in Lund, and the few traces that have been observed indicate that ordinary rectangular coffins were held together with either dowels or nails. Unlike the slightly younger graves in Lund, pagan burial traditions are reflected at Vannhög, for instance in the way graves were marked. Some have a mound over the grave, an honour that was probably reserved for people of high status in the village. In the excavated part of the cemetery, traces of six burial mounds were noted. Unfortunately, only one of them contained a preserved skeleton, the mound furthest to the east. An area that is otherwise very interesting is the one where three of the superimposed burials were found. Superimposed burials are ones where two individuals were interred one on top of the other with a certain time interval. This means that the

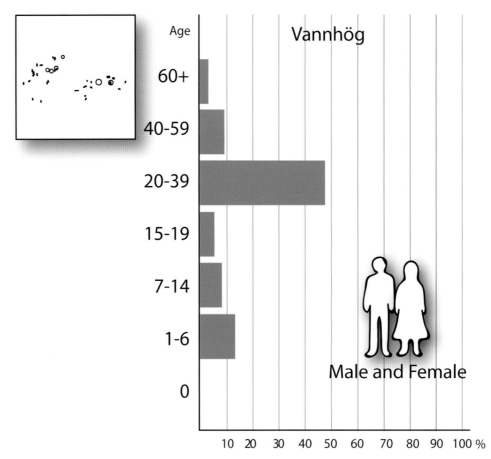

Fig 10 Age distribution at Vannhög. No infants found, but the number of children aged 1–6 and 7–14 is what can be expected.

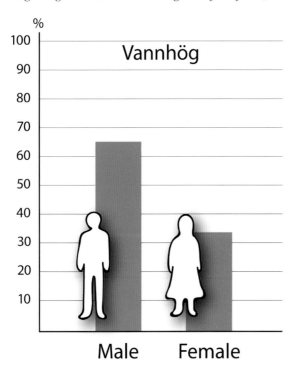

Fig 11 Sex distribution at Vannhög, which is skewed, with twice as many men as women.

site of the grave must have been so well marked that it was possible to place the next individual more or less directly on top of the previous body. At Vannhög there are four such superimposed burials. One of them is in one of the burial mounds, two are near it, and one is in the western part of the burial ground. It is also worth noting that the mound covering one of the superimposed burials cannot have been built until individual number two was buried. One of the superimposed graves found near the mound in the east contained two men (grave 200 and 307) that died at more or less the same age and they were buried with a knife and a bone needle. The man in the bottom of the grave also had a ceramic vessel (Fig. 12). Two big stones, one still in place, covered the buried men.

The position of the dead in the graves varied considerably. Some were laid on their back, others on the side, and one individual (grave 7), a child aged 7–8, has the legs drawn up so far as to give the impression that the child was tied up or died in an extremely hunched position and was buried soon after death. In the range of different body positions there is one that was not regarded as being equally natural and accepted by the society of the time, namely, placing the body on its stomach with the face to the ground. The man in grave 2 was found in this position.

Fig 12a–b Example of one of the superimposed burials at Vannhög. The buried were both men. A large stone was laid over the lower part of their bodies. (Photo Bengt Jacobsson Riksantikvarieämbetet UV Syd)

As was customary for non-Christians, several of the individuals were buried with both personal objects and garments belonging to the dress. The graves at Vannhög were not that rich regarding artefacts but contain finds like beads, a comb, a key, a knife, whetstones, spindle whorls of different size, needle holders, bone needles, and a vessel (Fig. 13). The knife was a very common object in the graves of both men and women.

Fjälkinge: a remarkable burial ground on the fertile plain

In the north-eastern part of Skåne, about 8 km east of the city of Kristianstad, lies the village of Fjälkinge. Rising in the northern part of the village is the natural formation known as Fjälkinge Backe (Fjälkinge Hill). The hill is elevated roughly 100 m above the plain and gives the visitor a beautiful panorama. People must have climbed the hill in the Viking Age. Perhaps they simply went there to get a good general view in order to decide on a good area to stake out as a burial ground. Today the cemetery is completely covered by residences and only the initiated know where the individual graves were originally located (Fig. 14). Just as at Vannhög, finds in the graves and radiocarbon dates together give a dating from the end of the 8th century to the start of the 11th century. And as at Vannhög, we do not know where the boundaries of this burial place were. Judging by the topography, the burials could have continued to the west, but to the east, south, and north there are natural limits in the form of a small brook and the sloping of the land.

In Fjälkinge the preservation conditions for wood are not as good as in Lund, but 47 graves out of the 121 excavated have stains left by coffins, and both adults and children were buried in them. Most of the coffins were rectangular. Only a few of them had nails, making it likely that the majority were held together by wooden pegs. Examination also revealed a couple of examples of trough-shaped coffins and showed that in two cases children were buried in iron-lined caskets with locks. Two of the graves (G 314, 399) can be certainly identified as double graves, in both cases for infants. In Fjälkinge we find six superimposed burials. In five of these, such a long time had passed between the burials that the body buried underneath was partly decomposed, with the result that parts of the skeleton were found in the filling of the burial above it. Three of the superimposed burials contained only infants (G 215, 360, 952), but in one of the graves (G 952) two dogs were buried in addition to the children; the dogs (one adult and one pup) were highest up in the stratigraphy. In the remaining superimposed burials we find an adult and a child in two cases (G 304 and 428)

Fig 13a–b Some of the artefacts found in graves at Vannhög: a vessel, spindle whorls, a key, a comb, bone needle and needle holder. (Photo Historical Museum, Lund)

and two adults in the other (G 65). The latter grave was first used to bury a very old woman. Her position in the grave was unusual: she was hunched up so tightly on her side that one can suspect that she was buried tied up or placed in a basket (Fig. 15a). The man, on the other hand, was stretched out on his back (Fig. 15b). They were both very old individuals, the man around 60 years of age and the woman 80 years of age. Maybe it was the mother and her son buried in the same grave but with some years in between. A pearl of amber was found in her grave and he had a knife and a comb. Yet another woman was found in this tied-up position, the individual in grave 349.

The position of the body also varies, and in Fjälkinge more than half were buried on their back, one on their stomach, one bent double, two tightly hunched up, and one with crossed legs. Of those buried on their back, 28 are children, three are teenagers, and 43 are adult individuals.

The remaining bodies were buried on their right or left side with slightly bent legs. Among the infants, the burial position could be determined in almost 80% of the graves, and among these half were buried on their side and half on their back. Just as at Vannhög the adult individual buried face down is a man, and he too was lying with his legs slightly bent (Fig. 16). The assessment of the archaeologist is that the pit that had been dug was not big enough and the body was carelessly dumped in it. No grave goods or personal finds accompanied the dead man in the grave. As for the man who was bent double, he was placed in a coffin that was big enough for his length but he had still been bent at the hips so that his head ended up on his knees. The absence of earth between the chest, head, and knees suggests that the man was placed in the coffin bent double, rather than the body being placed in seated position and later falling forward (Fig. 17). The grave gifts included a spindle whorl, some beads, a boar tusk, and a wooden bucket (only the metal handle was preserved).

The cemetery in Fjälkinge is judged to have been affected by the change of religion, an interpretation based on the fact that a number of graves are oriented north–south, several of them containing finds, while others are oriented east–west with the head to the west and are less often accompanied by grave goods. Those buried in a north–south direction with finds are regarded as typical of the pagan practice while the east–west examples lacking finds are influenced by Christianity.[29] An unusually large number of the burials are children, 88 out of 128 individuals (69%). Roughly a third of the infant graves show indications of Christian influence. The graves without finds are oriented east–west and the children are placed on their back. The pagan child graves are deeper and are oriented north–south. The children were laid on their side and a large number of them (28 out of 40: 70%) had pots placed in the grave (Fig. 18).

Just over half of the total number of individuals were given some kind of artefact in the grave. These could be personal belongings such as a knife or a comb. Some were accompanied by a bead necklace or a pendant with a Thor's hammer, made of either amber or metal. Brooches belonging to the clothing are evidence that at least three of the adults were buried in their clothes. Three of the graves were more richly furnished than the others. These are two adult graves, one man (G 600) and one woman (G 150), and a child grave (G 350). The man in grave 600 was buried in a lavish wooden coffin placed in a very large pit. The coffin was distinctive in having the ends retracted and the sides sticking out a little. There was no coffin like this in the cemetery. He was given a bronze buckle, two iron knives, and a whetstone. One of the knives was so big that it may have been a weapon rather than a tool. The belt buckle is also of high quality. The woman's grave contained two oval brooches and a trefoil brooch. The oval brooches are double-shelled, richly decorated, and gilded.[30]

Fig 14 Plan of excavated area and the graves of the cemetery of Fjälkinge, with the geographical location.

The richest grave in terms of the composition of finds, however, was grave 350, containing a child aged 2.5–3 years. The child lay in a coffin held together by at least 52 rivets. The grave goods consisted of a knife and a small whetstone, placed in the child's right hand. Around the neck was a necklace with two Thor's hammers, ten glass beads, 14 bronze beads, a cowrie shell, an amber pendant, and a carnelian bead; the latter is a precious stone with a reddish tone (Fig. 19). The cowrie shell does not occur in northern waters and must have come from somewhere around the

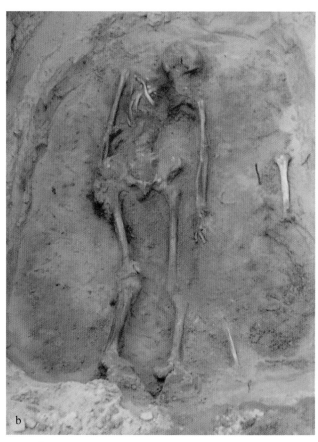

Fig 15a and b Individuals in one of the superimposed graves. The grave was first used to bury a very old woman around 80 years of age (15a). She was crouched tightly on her left side, possibly indicating that she was buried tied up or placed in a basket. The man, aged 60 years of age, on the other hand, was stretched out on his back (15b). (Photo Bertil Helgesson Kristianstad Museum)

Fig 16 Man buried in a prone position, lying with his legs slightly bent. (Photo Bertil Helgesson, Kristianstad Museum)

Fig 17 Man buried in a position where the upper part of the body was bent forward. (Photo Bertil Helgesson, Kristianstad Museum)

Mediterranean. The carnelian bead, with its pattern of polished facets, was likewise imported, possibly from the Caucasus or the Far East.[31]

The archaeological excavation and analysis of the material was done by Helgesson.[32] The osteological analysis was performed by Arcini.[33] The age distribution of the individuals buried in Fjälkinge is unusual in several respects. Unlike the other sites presented here, this one has an extremely high proportion of infants, while the share of older children is very small or virtually non-existent. At the same time, the proportion of older adults, especially older women, is high (Fig. 20). The sex distribution is equal (Fig. 21). Further details about the circumstances and underlying causes of the skewed distribution are discussed below in the chapter *Burial grounds designated for particular purposes?*.

Kopparsvik: a cemetery south of Visby

In the Baltic Sea, 100 km from the Swedish mainland, is the fabulous limestone island of Gotland. The light on the island is magical, and anyone arriving on the Gotland ferry on a summer evening at sunset will be enchanted. As you approach the island with the ferry from the west, you see the wall enclosing the medieval trading town of Visby, with its narrow alleys and old stone warehouses, a town for summer tourists from all over the world. When the Gotland ferry has berthed, you take the road that leads south, in towards the island, and on your left-hand side, below the cliff, you see some oil storage tanks. Here the view is not especially beautiful, but this was once the site of the large Viking Age cemetery of Kopparsvik (Fig. 22).

The Kopparsvik cemetery, with more than 300 excavated graves, is the biggest Viking Age cemetery on Gotland (Fig. 23). The site is just over 650 m from the wall that has surrounded Visby since the Middle Ages. Although the cemetery is not immediately adjacent to the natural harbour in Visby, the distance is very short. There is debate about which type of society lay behind the burial ground at Kopparsvik; it has been suggested that it represents a trading place, or a garrison for the hird or some other armed force.[34] We may hope that comparisons with other Viking Age cemeteries on Gotland and the mainland, and beyond the island and Sweden's present-day borders, will shed new light on this. Based on the finds, the graves are thought to date from the start of the 10th century up to the start of the 12th century.[35] Radiocarbon dates, however, indicate that the cemetery may have come into use as early as the second half of the 8th century.

The graves, which were dug in the shingle, lie along the former shoreline. Some of them are in just one layer, others in up to three. The lowest graves are sometimes as deep as 1.8 m. The cemetery consists of a northern and a southern area which partly overlap, and they are judged to have come into existence at more or less the same time.[36] The graves are what we call flat-earth graves and they are constructed of the stone that was available naturally on the beach ridge, that is to say, limestone slabs of varying size as well as some of granite (Fig. 24). Roughly two-thirds of the graves have some form of capstone. In other graves wooden coffins were used instead. In about 15 graves there are traces of wood which may be vestiges of planks. In the southern part of the northern burial ground there is a greater density of graves (Fig. 23). The cemetery had both double

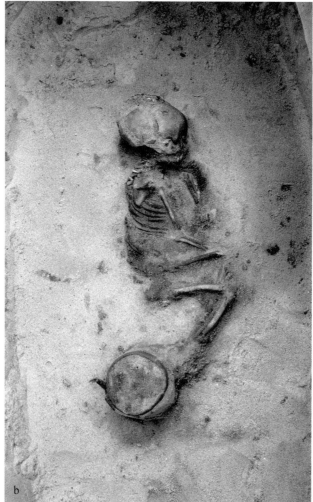

Fig 18a–b Two of the many small children's graves with the typical pot. (Photo Bertil Helgesson, Kristianstad Museum)

Fig 19a–b The best furnished and richest grave in the burial place at Fjälkinge, with among several things, a bead necklace with two Thor's hammers, ten glass beads, 14 bronze beads, a cowrie shell, an amber pendant, and a carnelian bead, a whetstone, a knife and a spindle whorl. Also depicted are some of the 52 nails from the coffin. (Historical Museum, Lund)

graves and superimposed burials containing both men and women, but with a predominance of men. Three of the four superimposed burials contain individuals in three layers. In these graves it was just as common to be buried on the stomach as on the back. This was not the case at either Vannhög or Fjälkinge, where the bodies were placed either on the side or on the back.

Burial face down during the Viking Age was a much more common custom on Gotland than in Skåne or Denmark. At Kopparsvik, no fewer than 12% of the bodies were in this position. One of the individuals at this cemetery had even been placed on his stomach with his lower legs bent straight up (Fig. 25). According to the archaeologist's interpretation, the grave shaft was probably too small and therefore the bones of the body were bent. Likewise, for those who were buried on their backs the graves had been made far too small, so that legs and feet ended up outside. The vast majority of these individuals were laid in a south-south-west to north-north-east direction with the head sometimes to the south-west. About ten individuals were buried in a west–east direction.

Roughly two-thirds of the bodies were buried with some form of personal belongings. The men's graves contained combs, strap buckles, strap-end mounts, ring brooches, and a large share of the graves contained knives, some an axe (Fig. 26a). Textile remains were found around brooches and pins, showing that the dead were buried in their clothes. Other finds from the graves were whetstones, fire-stones, combs, armlets, strap dividers, and a leather purse with bronze mounts. The women's grave goods are often linked to the garments they were wearing when buried, like bracelet, beads from a necklace, box-shaped and animal headed broches (Fig. 26b). Some box brooches of bronze, which are typical of Gotland, belonged to women of high standing in society, as shown by the silver and gold plating. Also found in the graves are beads and combs, sometimes with cases. In an impressive woman's grave, grave 112, Arabian silver coins were also scattered over the body. The coins are dated to AD 894–936.[37]

It is rare, however, to find weapons in the graves; only in four cases were there *scramasaxes* (one-edged long knives or short swords) in leather sheaths where the hanging systems were decorated, tin plated, and gilded. One of the men who was buried with a *scramasax* was laid on a bearskin (Fig. 27). Other weapons occurring in the graves were two spearheads and three axes.[38] The man in grave 50 was buried with scales of bronze, a number of weights, and an iron lock. If the grave goods are associated with what the man did during his life, it is conceivable that he was a merchant.

At Kopparsvik a total of 326 graves were excavated, containing 333 individuals plus further skeletal parts from damaged graves. There were probably other graves besides these which were destroyed by digging for marl, of which there is evidence in an area in between the northern and the southern parts of the cemetery. In contrast to the situation at the Fjälkinge cemetery, there were skeletons of only two infants/foetuses here, both of which were found in graves with women. Otherwise there was only one older child, along with a small number of teenagers in the age group 14–19 (Fig. 28). The sex distribution in the total material is highly uneven, with women accounting for only a quarter of the buried population. In the northern part of the cemetery the proportions are extremely skewed, with men constituting 8% of the burials, whereas the distribution is fully equal in the southern part (Fig. 29). A small proportion of the material represents graves excavated in 1956, about which there is some uncertainty concerning the exact location. It is clear,

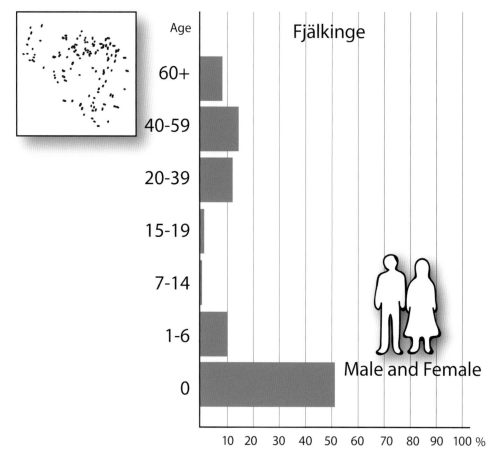

Fig 20 Age distribution at Fjälkinge. Note the extremely high proportion of infants, while the proportion of older children is very small or virtually non-existent. At the same time, the proportion of older adults is very high.

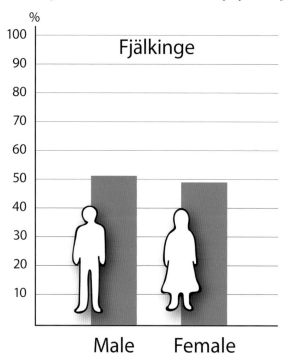

Fig 21 Sex distribution at Fjälkinge which is homogenous.

however, that they belong to the area between the northern and the southern section, but it has not been established which they belonged to. Most likely, however, they belong to the northern group of graves since the sex distribution is similar to that, with 78% men.

The excavation and analysis of the graves took place under the leadership of various archaeologists on different occasions from 1917 to 1964.[39] An extensive compilation of the grave material has been done by Thunmark-Nylén[40] and Toplak.[41] The osteological analysis on which this study is based was carried out by Arcini and Drenzel, and some of the findings have been compared with earlier analyses by Larje.[42]

Slite Square: with a view of sailing routes to the east

Slite is the fourth biggest community on Gotland in terms of population. In 1917 quarrying of limestone was started on the outskirts of Slite for the manufacture of cement. Today the quarry is a huge crater and the factory in Slite is one of Europe's largest producers of cement. At the site of what is now the square in Slite there used to be a large burial ground which was excavated several times during the 20th

Fig 22 Cemetery at Kopparsvik during the excavation. The location is very near the Baltic Sea. (Photo ATA, Riksantikvarieämbetet)

century. A total of over 50 inhumation graves have been found here (Fig. 30). The skeletons can give us information about what life was like in this area during the Viking Age. Slite is located in the bay of Bogeviken which, together with Visby, Paviken, and Fröjel was probably one of Gotland's major harbours in the period.

Finds in the graves and radiocarbon dates indicate that this cemetery was in use at roughly the same time as Kopparsvik, that is, from the 10th century and for 200 years until the start of the 12th century. Just as at Kopparsvik, the graves were made with material available on the site. The cemetery has not been totally excavated; there may be more graves to the south and west. The graves often had simple stone packings on the ground surface. These were usually irregular, in certain cases oval, rectangular, or triangular.[43] The body was usually placed on its back, occasionally on its side. None were found buried face down. As regards orientation, only one was east–west with the head to the west, in other words, with possible Christian influence. The bodies were placed on beds of gravel, and in eight cases wooden coffins could be identified. There were also structures of limestone slabs. One of the graves was a double grave, two were superimposed burials, and another two are assumed also to have been graves where individuals were laid layer upon layer. One of the superimposed burials contained two pregnant women, the lower one of which was placed on a layer of charcoal (47 (2A)).

Many of the bodies were buried with various types of artefacts, chiefly costume accessories but also jewellery. Examples of other objects deposited in the graves are needle cases with needles, tweezers, spindle whorls, but also equestrian equipment and household utensils such as pots and casks. One of the men (grave 8) was found with the remains of a money bag of a type that is believed to have its origin in the east. A similar bag has been found in the parish of Auster on Gotland, two in Birka, both in chamber graves (G 949 and G 958), and yet another one in Norra Åbyggeby, Hille parish, in Gästrikland.[44]

Unlike Kopparsvik, a larger proportion of the men's graves (5/14: 36%) contained weapons such as broadaxes, bearded axes, spearheads, and *scramasaxes*. In one grave there were also fragments of chain mail. Two graves had double weapon gear, containing both axe and spearhead.

A total of over 50 graves were excavated here, of which 43 contained skeletons. Osteological analysis has been performed on 30 individuals. Two of them are foetuses found with women. Otherwise there was one older child but no teenagers (Fig. 31). The remaining individuals are adults. The sex distribution is skewed in that 80 per cent of the adults are men (Fig. 32).

The graves have been excavated on several occasions by different archaeologists.[45] The osteological results are based on analyses by Mortágua.[46]

Fröjel: a burial ground beside a Viking Age harbour

If you drive south from Visby along the coast on road number 140 you come to the small village of Fröjel, large fields of poppies billow in the breeze, and the narrow roads are lined with viper's bugloss. On the right-hand side, between the road and the sea, stands the medieval church,

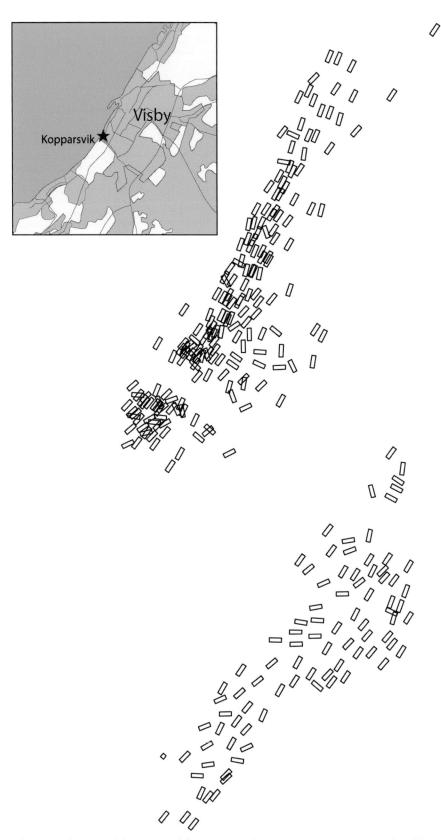

Fig 23 Plan of excavated area and the graves of the cemetery of Kopparsvik and the geographical location in Visby.

Fig 24a–b Graves at Kopparsvik are constructed of the stone that was available on the beach ridge, for example limestone slabs of varying size as well as stones of granite. (Photo ATA, Riksantikvarieämbetet)

Fig 25 Prone burial from Kopparsvik (grave 126). The lower legs are bent straight up. (Photo ATA, Riksantikvarieämbetet)

Fig 26a Examples of artefacts from male graves; combs, strap buckles, strap-end mounts, ring brooches and weights. 26b Female graves; bracelet, beads from a necklace, box-shaped and animal headed broches. (Photo Jan Nyborg, Gotlands Fornsal)

Fig 27 Male buried with a scramasax from Kopparsvik. (Photo ATA, Riksantikvarieämbetet)

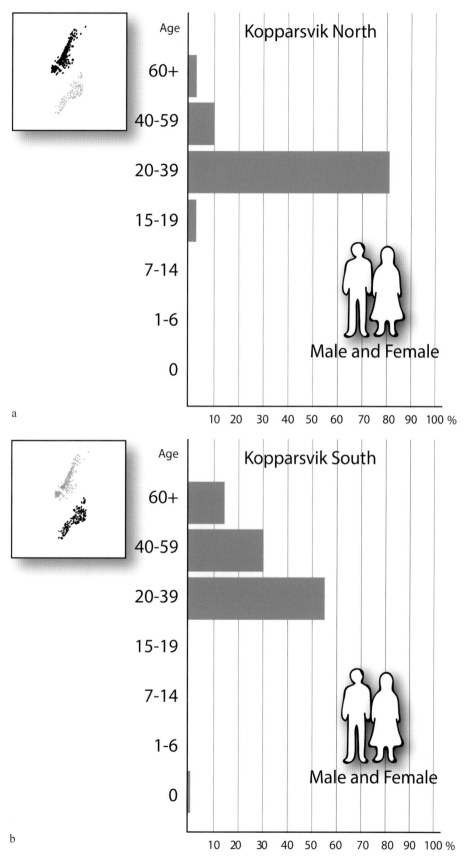

Fig 28a–b Age distributions at the northern and southern parts of Kopparsvik cemetery. Note that there are no children except for the foetus/newborn individuals at the southern part of the cemetery, and that there are more young adults at the northern part.

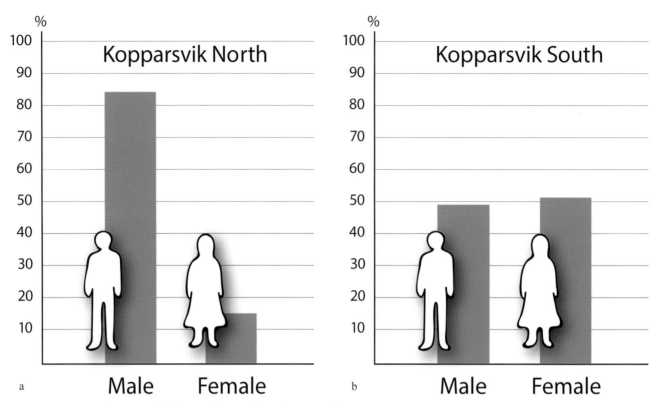

Fig 29a–b Sex distribution at both the northern and southern part of the cemetery at Kopparsvik, with a dominance of male burials in the northern part.

and just south-west of the church archaeologists have conducted several excavations of two areas with graves from the Viking Age and other periods, as well as areas with finds that can reasonably be associated with a harbour and boatbuilding.

Those who sailed to and from Fröjel could aim for the small bird islands Stora and Lilla Karlsö just south-west of Fröjel. On these islands guillemot and razorbill breed in large flocks. They may have done so back in the Viking Age too, and people from Fröjel may well have gone there to collect eggs. Fröjel, then, is a cemetery linked to a harbour, and not just any harbour. According to Dan Carlsson, who has done extensive studies of harbours on Gotland, Fröjel is one of the most important harbours on the island.[47] The place is about 40 km south of Visby. Archaeological excavations in the area have shown that there was extensive settlement here and two cemeteries, and also parts of an early Christian graveyard (Fig. 33). Judging by the finds, some of the graves here date back to the 7th century, but most are from the Viking Age. The burial ground consists of a northern and a southern part, neither of which has been totally excavated. Just as at Kopparsvik and Slite, stone is the predominant construction material of the graves in Fröjel.

Besides the excavated graves, an extremely large assemblage of stray finds indicates that they probably come from destroyed graves, meaning that the original number of graves in the area was larger. The finds from the excavations also suggest that the site was used for craft work. Weights for scales have been found here, as well as German coins, combs of different shapes, finished, half-finished, and blanks, and items of jewellery and the moulds for them. Also found on the site were objects such as nails, rivets, rivet washers, a rivet puller, and a ball hammer. The rivet puller was a tool for removing rivets from boat planks, and a ball hammer was used for the actual riveting, finds indicating that ships may have been built and repaired here.[48]

By far the most common finds on the settlement site at Fröjel, however, are potsherds, most of them coarse, undecorated everyday ware, but also a large amount of decorated pottery of the type known as Baltic ware. The burial grounds at Fröjel, as at Kopparsvik, consist of two parts, north and south. The northern area has both inhumation graves and cremation graves, dated from the 7th to the 9th century, the majority from the Viking Age. The southern area, by contrast, has only inhumation graves.

The osteological analysis of the 80 individuals shows that four were infants, three of them newborn (Fig. 34). In one grave the skeletons of two infants were found, one of them on or under the woman's right thigh bone and the other alongside the right foot. It has been suggested that they may be twins and that one of them, the one by the woman's

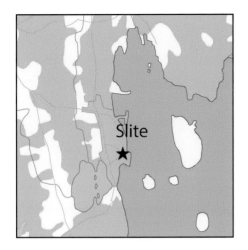
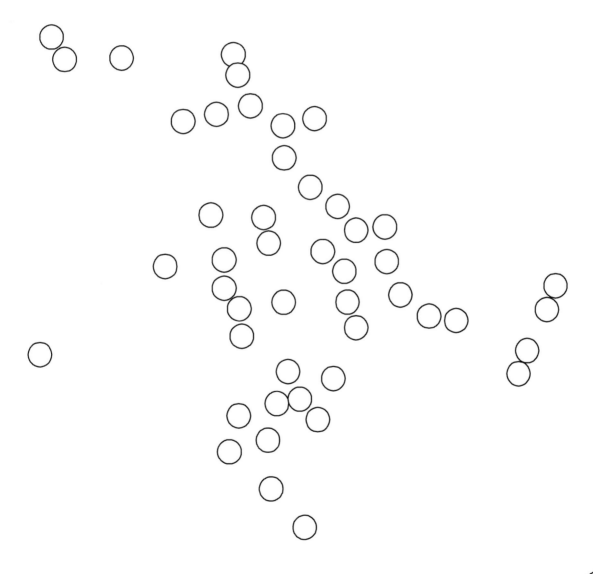

Fig 30 Plan of excavated area and the graves of the cemetery of Slite torg and the geographical location.

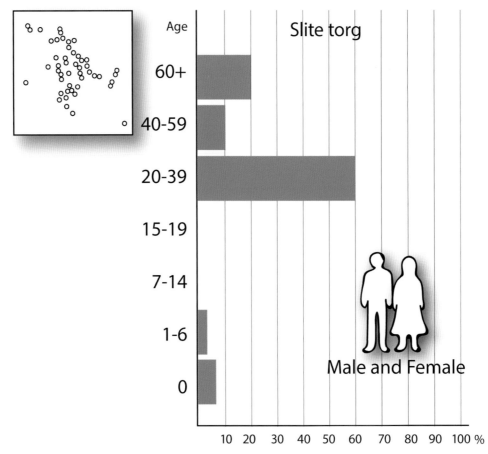

Fig 31 Age distribution at Slite torg, like Kopparsvik with most of the individuals being aged as adults.

foot, survived for a time and was buried later. This is highly improbable, however, since the bones of the children appear to be equally old and the ligaments in the woman's foot bones ought to have kept the bones in place for at least a year. It is possible, however, that the child buried at the foot is not the woman's but was nevertheless placed in her grave. The four infants were buried in the northern part of the burial ground. No older children were found in either the southern or the northern area, but there were eight teenagers, distributed in both areas. The remaining individuals are adults. Four of the graves were empty, with no assessment of whether they may have been graves for adult individuals or children. The sex distribution shows that the north part contrary to Kopparsvik is dominated by females (Fig. 35).

The excavations of the burial grounds, under Carlsson's leadership, took place on different occasions between 1987 and 1990.[49] The osteological analyses have likewise been performed on several occasions and by different osteologists.[50]

Birka: a well-known trading place in the realm of the Svear

To get to the island of Björkö in Lake Mälaren you have to go by boat. Here on the island is Birka, a place that has

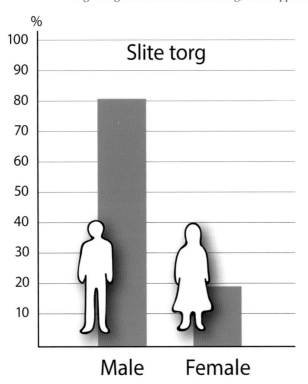

Fig 32 Sex distribution at Slite torg, with a male dominance.

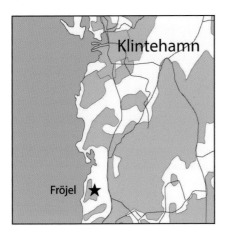
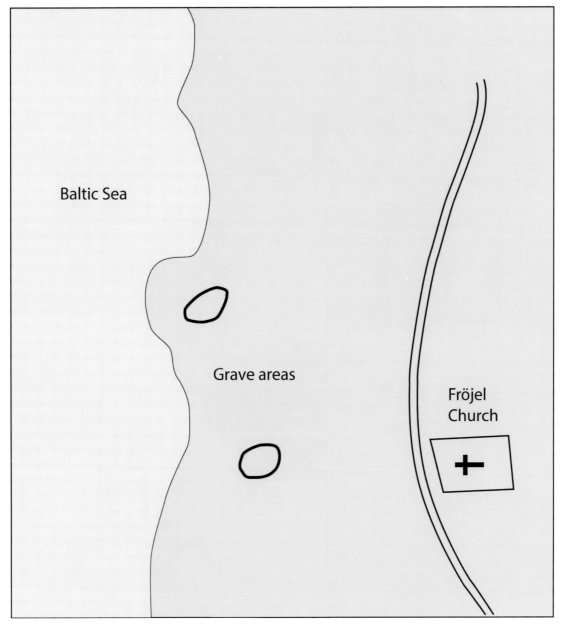

Fig 33 Plan of excavated area of the cemetery of Fröjel and the geographical location.

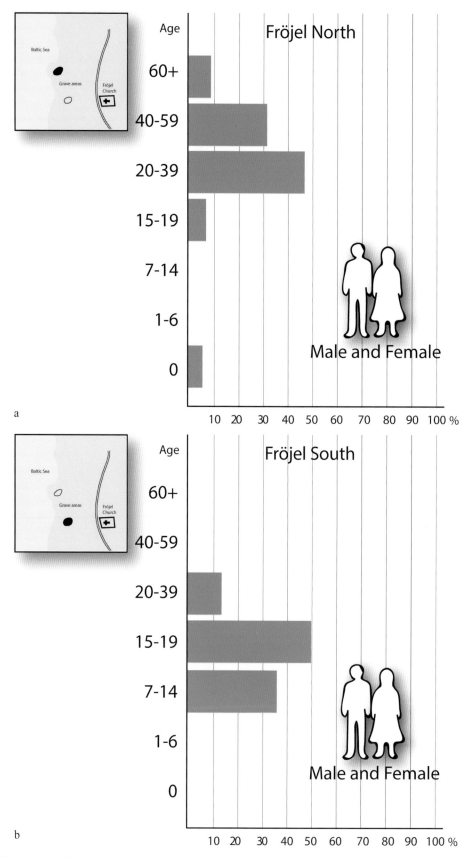

Fig 34a and b Age distribution of the two cemeteries at Fröjel. Comparable to Kopparsvik and Slite torg, with a relatively low proportion of children.

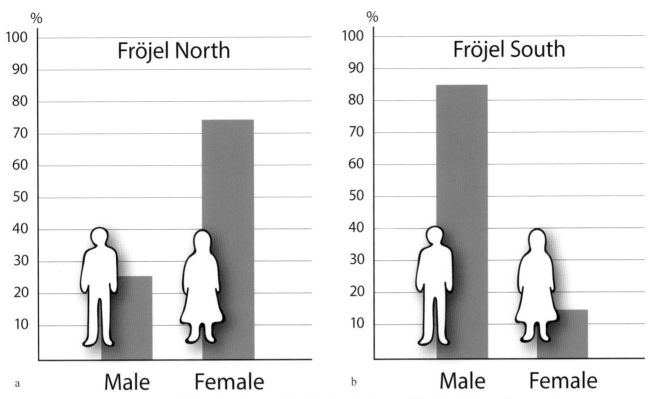

Fig 35a and b Sex distribution at Fröjel with a dominance of females in the northern cemetery.

thousands of visitors during the bright Nordic summer months. Birka has been on UNESCO's World Heritage List since 1993 and is one of the internationally best-known Viking Age burial and trading sites in Scandinavia. For over 150 years archaeologists have taken a keen interest in the traces of Viking Age life concealed on the island. Burial mounds, large and small, with or without standing stones, bear witness that many Viking Age graves still hide their secrets. At the end of the 20th century there were large-scale excavations of settlement and craft areas at Birka. To make it easier to understand what it looked like, some buildings have been reconstructed, and there is also a model of parts of the ancient trading site.

Birka was founded in the middle of the 8th century and was in use until *ca.* 975. Way back in the 1680s, Johan Hadorph, the 7th custodian of national antiquities, conducted excavations on the site and, in the 1870s, Hjalmar Stolpe excavated both inhumation and cremation graves.[51] Less than 100 years later it was time once again to dig in the black earth, this time under the leadership of Björn Ambrosiani. The excavations in the 1980s mostly concerned the town-like settlement.[52] Parts of what is called the Garrison were also excavated, an area outside the rampart that was Birka's defence system.[53] The extensive processing and publication of the graves from Birka was the work of Holger Arbman in 1933 and 1943.[54] The town or trading site flourished for 200 years but was abandoned at the end of the 10th century when the town of Sigtuna in Uppland was founded.

The only more or less contemporary written source that mentions Birka is the *Vita Anskarii*, composed by Archbishop Rimbert around 870. It is about the missionary Ansgar who visited Birka twice. Finds show that people at Birka in the 9th century had contacts, above all, with neighbouring regions and also with the kingdom of the Franks and with the western Slavs.[55] A hundred years later they were looking to the east, when contacts were established with Russia, the areas around the Black Sea, and the Caspian Sea.[56] The graves contain finds which indicate that long-distance trade was pursued with east and west alike.[57] Besides trade, there was much craft work at Birka, especially the manufacture of weapons and the equipment needed by a warrior.

People in general probably imagine that the graves at Birka are assembled in one big cemetery, but well-informed people know that the graves are divided among several different areas with different burial practices (Fig. 36). All in all, graves have been observed at eight different places on the island of Björkö.[58] Previous surveys have indicated that there were at least 3000 graves, but with modern technology we can say that the number is much larger, up to 4000–5000 graves.[59] A total of 1100 graves have been excavated.[60] According to Arbman[61] the number of excavated inhumation graves is 530. Unfortunately, preservation conditions for bone are poor, and it is only in 243 graves that remains of skeletons can be used for osteological analysis.[62] In addition a smaller number of graves containing skeletons have been excavated since then.[63] The burial grounds belonging to Birka that have been excavated are called Hemlanden,

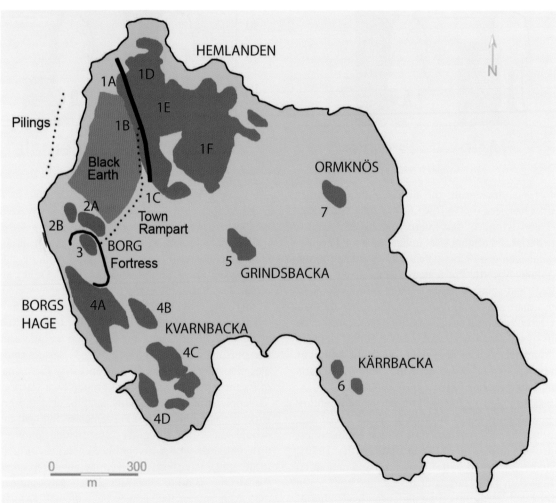

Fig 36 Plan of Björkö with Birka (grey) and the associated cemeteries.

Borgs Hage, Salviksgropen, Grindsbacka, Kvarnbacka, Kärrbacka Långnabba Puss, and Ormknös. Some of them may belong together, and Salviksgropen is called Ormknös B. The majority of the excavated inhumation graves come from the cemetery called Hemlanden (Fig. 36).

The burials at Birka are in both flat-earth graves and under mounds. Most of the bodies were buried in coffins, slightly fewer without coffins, and just over 100 in chamber graves. The term chamber grave refers to a wooden structure built on the site to contain not only the dead body but also, in some cases, a horse. The chamber graves also contain all manner of objects such as wooden buckets, crates, and weapons.[64] Chamber graves are found only at Hemlanden and the cemetery north of Borg.[65] The majority, 85%, of the bodies were laid

Fig 37 Burial ground at Birka. (The Swedish History Museum, Stockholm)

in a west–east direction. Apart from the fact that the skeletal material is poorly preserved, it is unfortunately also the case that there have been mix-ups over the years, because of which the number written on the bone bag does not always agree with the number of the grave. It is only for a small number of graves that it is possible to link the bones to the finds in the graves.

As regards the position of the body in the grave, in those cases where skeletons are preserved it has been noted that the majority are placed on their backs, a few on their sides in a crouched position, and one on its stomach (grave 724). In several of the chamber graves it has been observed that the legs are in a position suggesting that they were buried seated.[66] Visitors to Birka today can see that it was common to be buried under a small pile of earth (Fig. 37). Just as at Vannhög, Fjälkinge, Kopparsvik, and Slite, there were both double and superimposed burials. Of the chamber graves, seven were superimposed burials. In some cases individual number two was not buried until after the bones of the previous burial had been pushed aside to make room; in other contexts there was earth between the individuals, indicating that they were buried on different occasions. It also occurs that a coffin with yet another individual was deposited in the existing chamber grave, or that three separate burials took place, one on top of the other. There are three examples of women being secondarily buried in the chamber grave. The chamber was then primarily constructed for a man, whose skeleton was moved to one side. Gräslund suggests that these might be family graves where the wife, the widow, was later buried in her husband's grave.[67]

Fig 38 Tabletop pieces of glass from grave 523 in Birka. (The Swedish History Museum, Stockholm)

As at the other burial grounds presented above, there are double graves at Birka, at least nine of them. The finds in the graves at Birka are very rich and varied, probably because the site was an important trading centre, for example tabletop pieces of glass from grave 523 in Birka. Kent Andersson writes in his book that their origin may be from the Frankish area but it cannot be excluded that they come from as far away as the eastern Mediterranean (Fig. 38).[68] Furs and iron could probably be offered in exchange for goods from different parts of Europe and western Asia.[69] For further details of finds in individual graves the reader is referred to the Birka portal at the Swedish History Museum.[70]

Osteological analyses of the grave material from Birka shows that there are individuals of all ages.[71] The sex distribution of the adults in the inhumation graves, judging by the finds, was roughly equal.[72] The same has been observed by the latest osteological examination of the skeletons for which sex criteria are available, just over 40% of the adult individuals.[73] Among the cremation graves, however, the sex distribution is uneven; artefacts in the graves suggest that there are twice as many women as men.[74]

Skämsta: a farm cemetery

Between Uppsala and Gävle today lies the little village of Skämsta in Tierp parish. In the early days of the industrial revolution in the second half of the 19th century it became important for Sweden to use raw materials from the whole country. Railway construction began. At that time our ancient monuments did not enjoy the protection they have today, and parts of the burial ground at Skämsta were damaged, partly by the building of a house for the lengthman. Just over 100 years later, in 1994, the railway was to be expanded to a double track, but this time the site was protected, and archaeological excavations resulted in the exposure of the graves so that the skeletons could tell their story.

During the Viking Age the farm and the village were the social and cultural environment where most people spent their lives. Burial places and dwelling places lay side by side on the farmed land, and the burial places showed that the farm was inhabited for a long time. The graves at Tierp, north of Uppsala in Uppland, presumably represent part of a farm cemetery, dated by finds to the end of the Viking Age, around AD 1000–1100.[75] Six inhumation graves were found here, one cremation grave, and part of an individual's skeleton that was found in one of the other graves (Fig. 39). Older observations indicate that the graves were initially covered with mounds, but nothing remained of these at the time of the excavation. All the bodies still *in situ* were oriented south-west to north-east, five with the head to the south-west. All were buried on their backs with the hands along the sides, and all were accompanied by finds: combs, knives, beads, the handle of a wooden vessel, a large bronze ring too big to be a finger ring, and a small spirally ring of bronze judged to be a hair accessory (Fig. 40). According to the archaeologist, no counterpart to this hair ring has been found in Sweden, but there is a parallel in a Polish hoard.[76]

The osteological analysis of the eight individuals shows that the people buried included an infant, a large child/teenager, and six adults (Fig. 41). Among the adults there are three women, two men and one not sexed (Fig. 42). The boundary of the cemetery is not known. The graves were excavated by Frölund[77] and the osteological analysis of the skeletons was performed by Arcini.[78]

A wide range of burial practices

As we see from the descriptions of the different burial grounds, mortuary customs differed in Viking Age society. This is nothing new, of course; it has been known to archaeologists for over 100 years.[79] Variations can be seen both between and within the different burial places. The differences concern both the internal and the external construction of the grave, and also what the dead persons were given to accompany them to the next life. Of the sites included in this study, Birka, without doubt, displays the greatest variation as regards the outer and inner construction of the graves and the occurrence of finds. Although there are links to status, power, and thus the potential of the grave to stand out prominently, the external form is dependent on whatever construction material was available. This is most obvious from a comparison of, say, graves at Birka and graves on Gotland. On Gotland stone is generally available at many places and is therefore frequently used.

If we expose the construction inside the grave and compare the internal mortuary practice we find differences there too, from the simplest method of burying the body in a pit in the ground to the construction of an entire chamber of wood. The latter occurs frequently at the cemeteries of Birka, especially the westernmost parts of Hemlanden and even more commonly north of Borg and in and under the town rampart. Chamber graves are thought to belong to the upper stratum of the population, and the individuals buried there can be men, women, or children. Inspiration for this tradition is assumed to have come from Western Europe. As we shall see later in this book, strontium analyses of the people buried in these graves reveal interesting findings on that matter.

Other signs of high status and wealth are the finds present in the grave, when they are not Christian burials. Many graves with lavish finds occur at Birka, but the graves on Gotland are likewise richer in finds than those in Skåne. This does not mean, however, that the graves in Skåne are completely without objects. In Fjälkinge it is particularly clear that even a little child had a place in society; it is even the case that the grave with the most splendid finds was a child's grave. Several of the finds at the different burial grounds are not gifts but various details that were part of the dead person's clothes, as well as jewellery. At Vannhög the burial mounds, even though they are low, are a sign of status. The individuals in these graves, however, were not provided with any special objects. It is important to remember that several of the mounds are so damaged that both skeletons and artefacts are lost. At Fjälkinge it is above all the small child graves that contain finds. It is touching that people paid particular attention to really small children. The care that was taken is obvious in this group.

Another difference between the burial grounds is that both at Birka and at the Gotland cemeteries we find graves with weapons, a pattern that does not occur at the two sites in Skåne. Does this have something to do with cultural

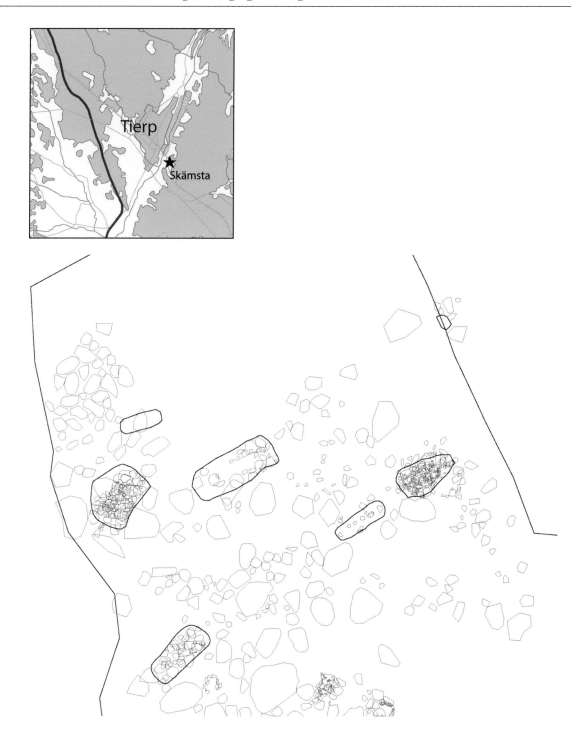

Fig 39 Plan of excavated area and the graves of the cemetery of Skämsta, with geographical location.

differences or the character of the sites, that is to say, the groups that used them? Are the differences due to varying cultural contacts with the surrounding world? These are questions that will be discussed later in the book when we have looked more closely at what the strontium analyses show, in other words, how large the foreign element at each place may have been.

At two of the Birka burial grounds – Grindsbacka and Kärrbacka – there are indications of Christian influence.[80] At Fjälkinge too we may suspect a Christian element in that the orientation of the graves changes to east–west and the graves contain fewer finds. Once we reach Christian times the position of the body also becomes more uniform, as the majority are laid supine. In the cemeteries, however, we find that, as long as the old religion still held people in its grip, the position of the corpse varied. The most remarkable position is flat on the stomach. In this study individuals have been found buried face down at five of the eight burial grounds: Vannhög, Fjälkinge, Kopparsvik, Fröjel, and Birka. Kopparsvik is the site where it was by far the most common, accounting for 10–12% of the burials. The practice is by no means unique for the Viking Age or for Sweden, but in Scandinavia it is attested chiefly during the Viking Age.[81] The oldest known cases have been found in the present-day Czech Republic and they are 26,000 years old. The youngest is from the First World War.

There may have been various reasons for burying people on their stomachs.[82] It has been suggested that they were criminals and that their sins had to be signalled by burying them face down, as if the punishments they had received in life were not sufficient. This would explain the odd position of the arms and legs of these individuals, for example, the arms behind the body, indicating that they were tied. In some cases executed individuals have also been found buried face down, but the majority of those who had been beheaded or hanged were buried on their backs.

Other suggestions as to why individuals may have been buried on their stomachs are that they died outdoors in this position and the face was not in a fit state to be seen, and therefore the body was turned. It has also been speculated that it could have been people who drowned and floated ashore. Other proposals are that the individuals buried on

Fig 40 Artefacts from the graves at Skämsta; a comb, knife, beads, a large bronze rings. The spirally ring of bronze probably represents a hair accessory.

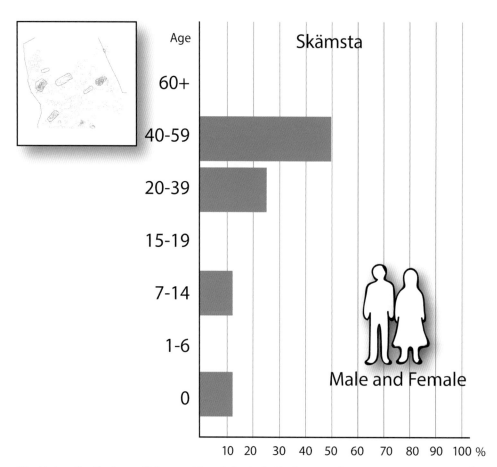

Fig 41 Age distribution at Skämsta, although few individuals, several age groups are represented.

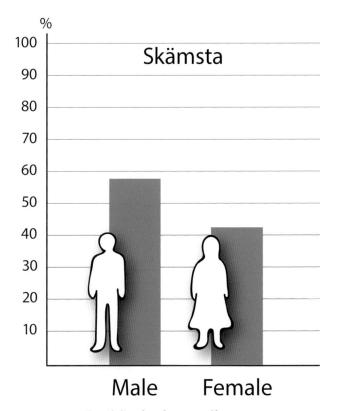

Fig 42 Sex distribution at Skämsta.

their stomachs were shamans whose power people feared. That sick people could have been treated in this exceptional way is another interpretation. A simpler explanation is that it is carelessness or a lack of concern for the individuals who were to be buried. Bodies buried in shrouds could have ended upside down because it was impossible to see which was the right way up. In cases where the body was buried in a coffin, the suggested explanation is that the person was unintentionally buried alive and turned in the coffin in the hope of forcing his/her way out.

There is little written evidence about the burial of people face down. One written source from medieval France, however, indicates that even important kings at the beginning of the period we call the Viking Age had their kinsfolk buried this way. In Abbot Suger's description of the rebuilding of the church at St Denis we read that the work started with the older entrance, which was said to have been built by Charlemagne (742/747–814) for a special reason, namely because his father Pippin III (the Small) (714–768), king of the Franks 751–768, had asked to be buried outside this church face down, to atone for the sins of his father, Charles Martell (686–741).[83] The common denominator in several of the different explanations for the face-down burials is that they all signal a treatment of a fellow human being who was no longer deserving of respect. The global study indicates that among those who were buried on their stomachs there were men, women, and children, but the men dominate. The custom is not connected to any specific culture or religion and it is represented all over the world.[84]

Mattias Toplak, who has done in-depth studies of Kopparsvik, has a different suggestion, chiefly connected to the conversion to Christianity. He believes that it is not intended at all as degrading treatment, but that the probable answer can be found in the written sources, reflecting the ritual of "prime-signing", that is, the first step to baptism.[85]

Everyone was buried

The overall demographic picture we obtain from the different burial places shows that individuals of all ages were interred, with nothing to suggest that any age group was excluded, and with both sexes at every burial place. What the study clearly indicates, however, is that distributions are skewed for various reasons. Demographic studies based on archaeological material often find that certain age groups dominate and that some are poorly represented or wholly absent. This study is no exception and, as so often, infants and old people are the age groups with the lowest representation. At the same time, this survey shows that both infants and old people are well represented at several places (Figs 6, 10, 20, 28, 31, 34, and 41).

Previous studies have demonstrated that the representation of men and women can vary between different burial sites, and this is also the case here. The general picture in this study is that men in total are more numerous than women (Figs 7, 11, 21, 29, 32, 35, and 42), although two of the burial grounds, Birka and Fröjel, show the opposite. The greatest disparity in sex distribution has been observed at Kopparsvik.

What is the reason for the skewed distribution? In this chapter the aim is merely to display the variation. A more detailed discussion of the underlying causes will come in the chapter *Burial grounds designated for particular purposes?*, where different parameters will be considered as an aid to the assessment of the type of sites we are dealing with.

3

Immigrants or locals?

Who were the people laid to rest in the graves? Did the majority come from the site, or could foreigners be buried there too? The Viking Age was a time in prehistory when the Nordic populations widened their horizons and travelled far beyond the Scandinavian peninsula. This was a time when people travelled in all directions and settled in both uninhabited and inhabited places. Long-distance trading contacts, however, had started even before the time we call the Viking Age, contacts that were now expanded, and not always by peaceful means. Scandinavians had realized the wealth that could be gained abroad. New lands were colonized and conquered. Existing populations were subjugated, but alliances and contacts were also established between strangers at both high and low levels. This resulted in connections with other cultures and religions which would influence the Scandinavian population for all time. Inscriptions in the form of runes found at different places in Europe, eastern and western chronicles, and the Icelandic sagas are part of the foundation for our knowledge of where Viking Age people were taken by their curiosity, ambition, greed, and skill. Apart from these sources, Scandinavian artefacts of the time that have been found across Europe, and exotic foreign objects in Scandinavia show that it was a society with a wide geographical network of contacts.[1]

From Denmark as it is today, and perhaps also as it was then, including today's Skåne, people set off chiefly to England but also southwards to Francia, the Iberian peninsula, and northern Africa. Those who lived within the boundaries of present-day Norway headed not just to the British Isles but also north-west, across the Atlantic to the Faroes, Iceland, and Greenland, and even as far as North America, while those who lived in the eastern part of Sweden, north of Skåne, tended more to move eastwards across Russia on rivers like the Volga and the Dniepr towards the Black Sea and the Caspian Sea, down to the Byzantine Empire in today's Greece and Turkey (Fig. 43). Finds of different kinds and statements in written sources reveal that contacts were forged as far away as present-day Iran, Iraq, Afghanistan, and Uzbekistan.[2]

The reason for the Norse expansion and raiding is thought today to be the quest for wealth, more so than emigration as a necessity. The older thesis that population growth impelled migration is thought today to be valid only for western Norway.[3] The piracy of the Norse as described in the chronicles is judged to have begun 100 years before the raids of the Viking Age proper. People in the Scandinavian peninsula and the lands around the Baltic Sea had become aware, via trade, of the wealth that could be found on the continent and in the British Isles. It was during the 8th century that trading sites like Dorestad, London, and York were established, and the Norse traded in goods such as furs, eider down, amber, and high-quality whetstones.[4] That was also when trading centres like Ribe, Hedeby, Birka, and Staraja (Old) Ladoga arose.[5] To what extent these early contacts meant that the Norse themselves travelled or if it was mainly trade with people visiting Scandinavia is difficult to say.

At this time much of Europe consisted of many diverse chiefdoms which were already in conflict before the Norse entered the picture. The pagan Norse took advantage of these conflicts by taking sides, first with one and then with the other, and keeping the peace in return for large sums of money.[6] In the chronicles of Western Europe the Vikings have acquired a reputation for being extremely destructive and ruthless but a new, more nuanced picture indicates that they were no more brutal than the groups they attacked. The aim of the Vikings was usually to plunder and take prisoners whom they later released for large ransoms. In Ireland,

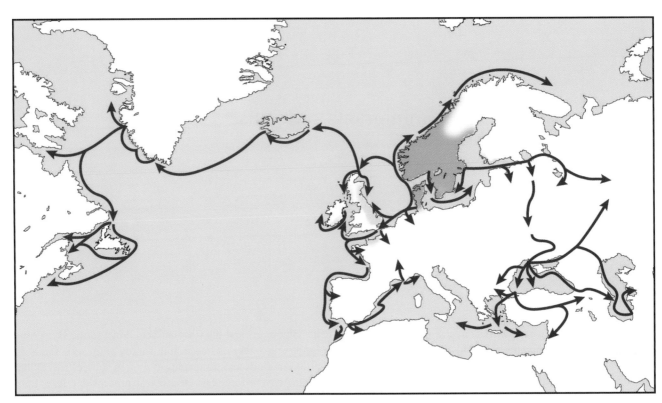

Fig 43 Map with Viking Age routes, AD 750–1100. (after Hjardar and Vike 2011)

by contrast, the churches were poorer and therefore the Vikings did more slave trading with their captives, but the most extensive slave trade was pursued in Eastern Europe, where slaves were sold to Muslims in exchange for silver.[7]

How large the migration from Scandinavia was – *how many* people moved away during the Viking Age – is unknown. How mobile people were within the Scandinavian peninsula and how many people moved here, on the other hand, is something we can have some knowledge of. From written sources we know that some people came to Scandinavia of their own free will, for example to preach the gospel or to pursue trade, while others came here by coercion.[8] Some came only on brief visits, others to stay. Even if it was not their intention to stay, things may have happened, for example, that they died while visiting and had to be buried somewhere. The question is, how did people handle the situation? Were foreigners buried in the local farm's burial ground? What customs were practised? Can we detect which people originally came from the place and which of them came from outside?

A geological signature can be detected in dental enamel

In the last 20 years, studies of stable isotopes have brought completely new opportunities for answering questions about mobility. Elements occur in different forms, known as isotopes. ^{87}Sr is an isotope of strontium formed through radioactive decay of the element rubidium (^{87}Rb). Rubidium has a long half-life (radioactive decay), which means that old bedrock has proportionately more ^{87}Sr. The content of strontium (^{87}Sr) thus varies depending on the age of the bedrock. Through food and water, we consume the strontium isotope, which is stored in dental enamel but also in the skeleton.[9] The geology of an area gives us a hint of what we can expect; if the bedrock is old it gives a high strontium value; young bedrock gives a low strontium value. The amount of strontium in bedrock is well known, however, it is not on the bedrock that people live, but on the loose soil cover, and this contains material that may have been transported far and is not always identical with the bedrock. One way to find out the strontium value shown by people from a particular place is to measure the dental enamel of animals found at the place. It is best to use rodents who do not move over large areas, but sometimes circumstances force us to select other types of animals. The values we obtain from the animals are used as a baseline against which we can compare values from the humans we want to analyze.

By measuring the ratio of strontium (^{87}Sr/^{86}Sr) in the enamel of permanent teeth, we can say whether an individual has moved from one geographical area to another, provided the two areas differ geologically. The method allows us primarily, and at best, to draw conclusions about which persons grew up in a place and which persons moved there, whereas it is harder to say where the incoming individuals came from. In recent years, a large number of strontium analyses have been performed on archaeological material from Viking Age Scandinavia.[10]

In this study Price has done strontium analysis on individuals from the Trinitatis cemetery, Vannhög, Fjälkinge, Kopparsvik, Slite, Birka, and two analyses of the burial place in Skämsta. To obtain even more data for Gotland we have also used results from analyses of the material from Fröjel, taken from a master's thesis by Pechel.[11] Other grave material is included for comparison and to provide a more solid basis to give us a general picture of mobility during this exciting period. This material comes from burial places on Öland;[12] Bornholm, Sjælland, and Fyn in Denmark; and Saaremaa in Estonia.[13]

As the map shows, the different cemeteries studied here are in places with partly different geological bedrock and ought therefore to display different strontium values (Fig. 44a–c; Appendix Table 1). Some of the places we study are in relatively homogeneous areas as regards geology, while others are located in areas with more heterogeneous conditions. To simplify matters, one can say that in areas within the boundaries of present-day Sweden we cannot expect values lower than 0.710. Local individuals from Gotland are judged to have values no higher than 0.713, in Skåne and on Öland the values are no higher than 0.717 and in Birka on Björkö the highest value is 0.728. However, Birka is in an area with varied geology and in the north the values can rise to 0.739 (Fig. 45a–c; Appendix Tables 1 and 2). A closer examination shows that Trinitatis in the town of Lund is on the boundary between different geological areas. As a town, of course, Lund was partly dependent on its hinterland for its food supply, while drinking water came from wells in the town. To assess which strontium value one expects from the local inhabitants of Lund and its immediate surroundings, two black rats from the town have been analysed.[14] To include parts of the hinterland, earlier results from a larger number of animals from the well-known Iron Age site of Uppåkra, less than five kilometres southwest of the town, have been used. In the diagram (Fig. 45a; Appendix Table 1) we see that roughly 80% of the values fall within the interval 0.711–0.712.[15] People who grew up in Lund and its immediate surroundings, at least south-west of the town, should thus display strontium values within the stated interval. If, on the other hand, they brought food from the south-east, towards Österlen, or in northern Skåne, we can expect slightly higher values (Fig. 44b; Appendix Table 1).

Vannhög in Trelleborg is located in a more homogeneous geological area. Here, unfortunately, we had access to only two animals, both dogs, but their values harmonized. They gave an interval of 0.710–0.711, which is reasonable given the age of the bedrock in the area (Fig. 45b). The material may be small, but it is all that we have at the moment. For the burial ground in Fjälkinge the conditions are problematic. In the area where the burial place is located, bedrock of different ages meets. The foundation on which

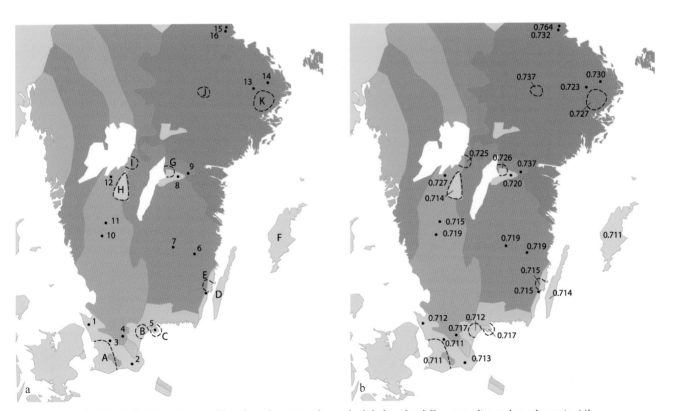

Fig 44a–b Geological map of Sweden where 44a shows the label to the different median values shown in 44b.

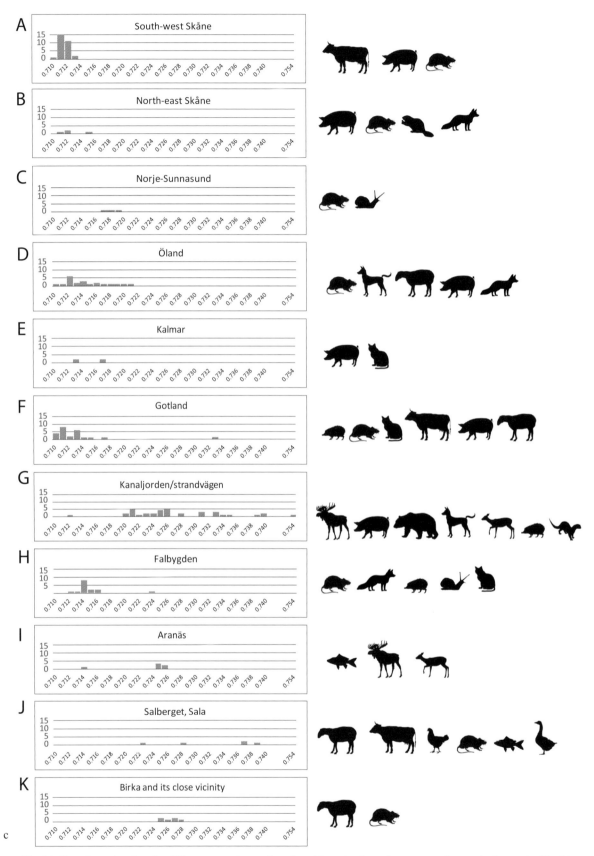

Fig 44c Bar chart showing the different strontium values found in animals and corresponding to the letters in Fig 44a.

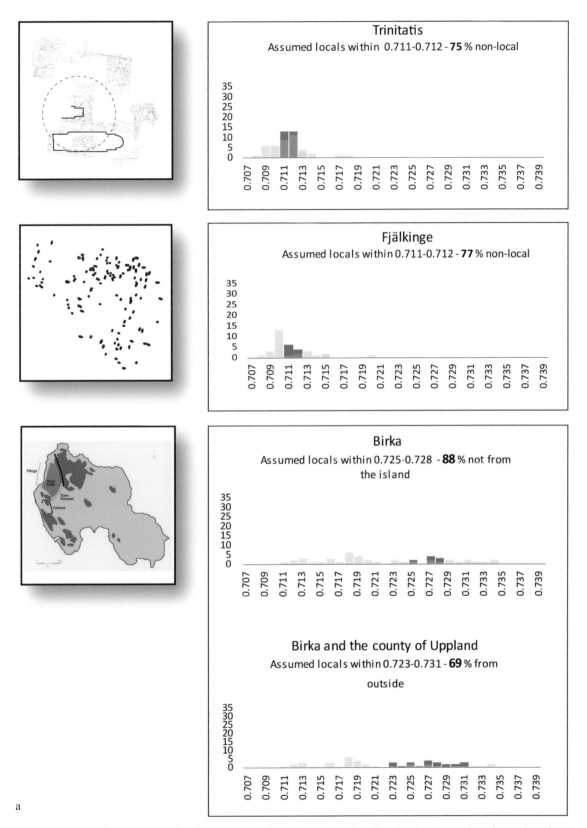

Fig 45a–c Representation of strontium results: the dark brown bars indicate the baseline, local signature based on values from animals, plants and water (majority are from animals see Table 1) while the light browns are judged as non-local. Similarly, the dark blue part of the stack represents the local people, and the light blue non-local individuals.

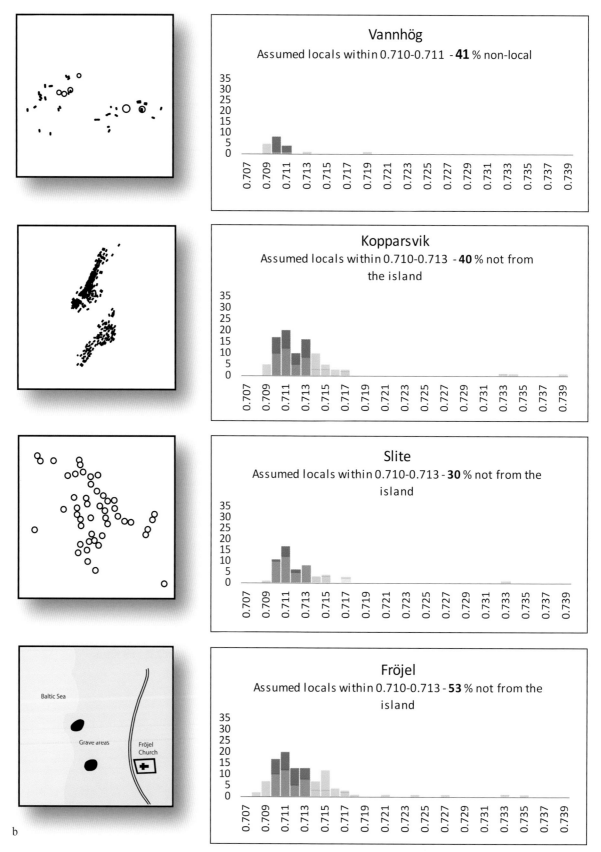

Fig 45a–c Representation of strontium results: the dark brown bars indicate the baseline, local signature based on values from animals, plants and water (majority are from animals see Table 1) while the light browns are judged as non-local. Similarly, the dark blue part of the stack represents the local people, and the light blue non-local individuals.

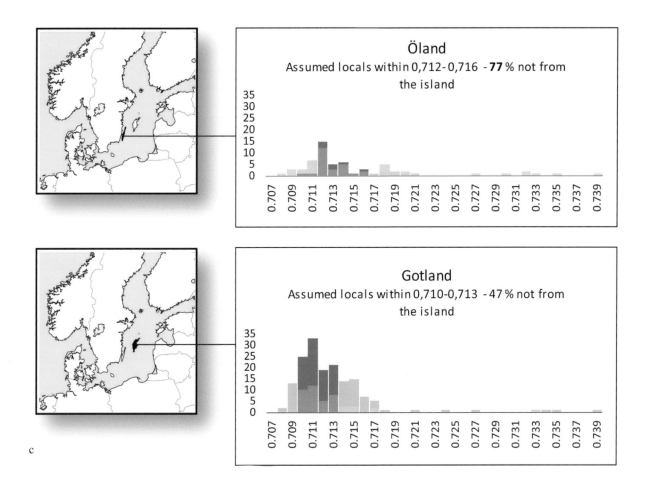

c

we determined the baseline consists of results from four animals, three wild and one domesticated. The diagram shows that three of the animals lie within a narrow interval, 0.711–0.712, while the fourth animal, a fox, indicates a higher value, 0.715. The explanation for the discrepant value for the fox may be that its territory is larger, 5–20 km, and it can thus be expected to have a higher strontium value. Our assessment is that the people who grew up in or near Fjälkinge ought to have had strontium values of 0.711–0.712 (Fig. 45a; Appendix Table 2). People with slightly higher strontium values 0.713–0.715 may have grown up within a somewhat larger radius of the actual burial ground.

As for the island of Gotland in the Baltic Sea, limestone is the predominant bedrock. Those who grew up and lived on what the island gave ought to show a fairly homogeneous and rather narrow range of strontium values. Analyses of both wild and domesticated animals found in the vicinity of the town of Visby, and samples of plants and water show that 80% of the strontium results fall within the interval 0.710–0.713 (Fig. 45b; Appendix Table 2). It is therefore probable that those who grew up on the island are also within the interval. Some animals from the Viking Age settlement site at Fröjel, however, gave higher values and one of them gave an extremely high value, which exceeds what can reasonably be expected on the basis of the geological conditions on the island (Fig. 45b; Appendix Tables 1 and 2). This animal, a pig, like the other three animals with higher values, is judged to have been imported to the island. It would not be strange to bring a live pig here from somewhere on the mainland. During the Viking Age animals were transported from far afield; for example, horses were shipped to Iceland and even further away, to Greenland, in connection with colonization.[16]

The last place we have to discuss concerning strontium is Birka. Birka was well known as a trading site, with everything that involves in contacts with the surrounding world, and it lay in an area with greatly varying geology within a relatively short distance (Fig. 45a; Appendix Table 2) This means that we can expect far more varied strontium values solely because people moved over relatively short distances. Analyses of animals from Birka, as we have seen, gave a narrow interval of 0.725–0.728, which ought also to be found in humans who grew up on the island (Fig. 45a; Appendix Table 2). Animals from other areas in the Mälaren valley and a few dozen kilometres north of Birka, however, show much greater variation, 0.723–0.731 so we can reckon with a larger interval here for those who had trading contacts with the island and who then settled there for a more or less long time (Fig. 45a; Appendix Table 2).[17]

The work of using strontium analyses to ascertain the values that can be expected at different places in Sweden has just begun. Our map thus has many blank spots. This study, however, contributes new results. For places with relatively homogeneous bedrock and hence rather narrow strontium intervals, such as Gotland or western and southern Skåne, it is therefore interesting to find out the origin of people whose values do not agree with those common in the area. Did they come from nearby areas or further away? Data from some of our closest neighbours also give us some knowledge about their geological conditions.

The choice of individuals for strontium analysis in this study has been determined by whether teeth are preserved or not. Although there are occasional children in the study, the results are mainly based on adults in the population. The study should primarily be viewed as a compilation of current knowledge aimed at gaining information about conditions in the Viking Age population. A total of 175 individuals have undergone strontium analysis and a further 90 individuals from comparative materials from Gotland and Öland. Other comparative material comes from Bornholm, Sjælland, and Fyn in Denmark, and the early Viking Age boat graves from Saaremaa in present-day Estonia are also included. The discussions concerning mobility during the Viking Age are thus based on 405 strontium analyses from several different places around the Baltic Sea.

Different patterns emerge

What then are the results of the strontium analyses? If we begin to paint a picture with a broad brush and make comparisons between the different burial grounds, certain patterns emerge. At the Trinitatis graveyard in Lund and the cemetery in Fjälkinge, as at Birka, the proportion of non-locals is very high (Fig. 45a). At Vannhög in Trelleborg, as at the three Gotlandic burial places Kopparsvik, Slite torg and Fröjel, the situation is different; the non-local element is smaller (Fig. 45b). A shared feature of five of the six sites, however, is that they contain groups of individuals with a common denominator, namely, lower strontium values than expected for Sweden, 0.708–0.709. They occur in the largest proportions at Trinitatis in Lund and Vannhög in Trelleborg, but smaller at Fjälkinge and the Gotlandic burial places of Kopparsvik, Slite and Fröjel (Fig. 45a–b). The results from Öland are roughly the same as those on Gotland (Fig. 45c). In Birka, by contrast, this group is not represented at all among the individuals who have been analysed (Fig. 45a).

Another interesting observation is that at the cemeteries on both Gotland and Öland we find the high strontium values 0.725 and upwards, corresponding to those in the Mälaren valley, whereas values of this type are absent in Skåne. However, there are indications in Skåne, through an example from further back in time, that contacts with these regions may have existed in occasional cases, for example,

the cemetery at Önsvala. One of the graves containing a woman, dated to the Migration Period/Vendel Period, had a value of 0.725. Moreover, beside the woman's fibula there was a collection of amulets which included a beaver tooth. This has undergone strontium analysis, giving a value of 0.726. The archaeologist Lars Larsson's interpretation is that the woman came to Skåne from central Sweden.[18]

Although we will never know the reasons why individuals moved, we can ask if those who came stayed together as a group or mixed with the local population. The clearest group has already been named, those with low strontium values (0.708–0.709) and the question is: can we detect from their mortuary customs or their placement in the burial grounds whether they still constituted a distinct group? At one of the sites a clear grouping has been noted, namely, at the Trinitatis graveyard in Lund. There we find the individuals with low values on the edges of the burial ground. As we have seen, the town of Lund was founded at the end of the Viking Age. According to several scholars, both historical sources and finds from the oldest phase of the town indicate that it was founded with the aid of people of varied geographic origin.[19] Our study of strontium strengthens this hypothesis.

But where did the individuals who populated the town come from? According to several scholars, the occurrence of large quantities of Baltic ware suggests that there was a noticeable Slavic element and that this group may have been moved from the south coast of the Baltic Sea, more specifically the area extending from around the River Warnow in the west to the estuary of the River Oder in the east (Fig. 46).[20] And in this geographical area strontium values of between 0.709–0.713 have been measured in animals.[21] Yet strontium values of 0.708–0.709 are also found in Denmark[22] and in south-eastern England (Fig. 47).[23] According to written sources and numismatic research, and judging by the craft products that have been excavated, the people who came from England to Lund were above all coiners, goldsmiths, and priests, who belonged to a stratum of society that we would rather expect to find among the graves closest to the church.[24] Another geographically close area with correspondingly low strontium values is Jutland, but why the individuals with low strontium values on the edges of the Trinitatis cemetery should have come from there to Lund is difficult to say; there are at any rate no indications of that in the find material. This group of individuals with lower strontium values is only present among those living during the establishment of Lund, not during the later time periods.[25]

What about elements of Slavic origin at the other sites? Do they likewise show a link between the presence of Slavic pottery and the occurrence of low strontium values, 0.708–0.709? The burial place at Vannhög was probably used by a number of farms over a long time. There is no Baltic ware at Vannhög, but the Slavic Feldberg ware has been found in the Trelleborg fort in Trelleborg, which suggests Slavic

Fig 46 In the dark grey area north of the river Elbe we find the Slavic ware whose design was a model for the locally produced Baltic ware that was found in Lund in large quantities.

contacts. Feldberg ware, unlike Baltic ware, is imported.[26] However, there are finds at Bjäresjö and Mölleholmen. At Fjälkinge small pots were deposited in a large number of child graves, but this pottery is of the indigenous Viking Age type and not influenced by people of Slavic origin. However, there are different finds of Baltic ware at a village settlement site at Fjälkinge.[27] Regarding Gotand and Öland there are also finds of Baltic ware but with origin from the east (Riga).[28,29]

That Scandinavians had early connections with the Slavs in the southern part of the Baltic already during the 8th century is shown by the Feltberg ware that has found in, for example, Trelleborg, Ystad and Åhus but also at Paviken Gotland.[30]

What other interesting observations and conclusions could be inferred about the original inhabitants and newcomers? The large cemetery of Kopparsvik on Gotland is a result of two burial grounds having grown together. In the northern part the sex distribution is highly uneven, being male-dominated. The reason for this is a matter of debate, and one of the questions has been whether the burials represent people from outside or an indigenous people. The number of strontium analyses is roughly proportionate to the total number of burials in both the northern and the southern parts,

Fig 47 Expected strontium values based on measurement of animals, plants and water in England, Norway, Denmark, Germany (Schleswig-Holstein) and Estonia (Salme).

and the results show that there are more local people in the northern area than the southern (Fig. 48). In the southern part we also find four of five individuals with low strontium values, while simultaneously the two with the highest values are buried there. The latter are moreover buried very close to each other. In other words, this suggests that the male-dominated burial area represents the local population rather than non-locals.

However, strontium values from 0.723 to 0.731 have been documented indicating that several of those buried at Birka were born and raised in the area of Uppland. Still, a large percentage of them, especially those with much lower strontium, were people coming from quite far away, from regions like Gotland, Öland, Skåne, and perhaps also Estonia (Fig. 45c). Five of the analysed skeletons from Birka represents individuals between 5 and 14 years of age and only one of them with a value corresponding to Birka, 0.728, while the other four show values indicating that they came from elsewhere, 0.718 (Appendix Table 2).[31] The question is whether undiscovered lower values could nevertheless be found near Birka or if their origin should be sought further away, that is to say, even the children came here from some other place.

Were there differences between men and women? This could inform us about post-marital residence. Was it more common for men to move, or do the result show that it was the women who came from outside? The answer is that it varies from place to place, but the general picture is that it was equally common for men and women to represent the non-local element (Fig. 49). On the other hand, lower strontium values (0.708–0.709) are more common among women than men (Fig. 49). A reasonable explanation could be that they are among those that represent the female ceramists that produced and introduced the Baltic ware to Scandinavia. According to Roslund, Baltic ware was made by Slavic women in Scandinavia, who continued their tradition in their new homes working as what may be referred to as slaves for the household. As has been mentioned before, individuals with strontium values higher than 0.720 have not been seen in burial places in Skåne. They are, however, found on both Öland and Gotland (however not in such a high degree as those with values

Fig 48 The cemetery of Kopparsvik where the light blue represents the local individuals while the dark blue indicates the non-locals.

(0.708–0.709) and among them there were a few more men (Fig. 45a–c). A suggestion for this could be that the islands of Öland and Gotland had transit harbours for trade from, for example, the regions of Mälaren valley or further north, or even traders from regions of Russia. Regarding Birka, 21 individuals of 37 (57%) have been sexed (nine women and 12 men) and among those we see a somewhat larger variation among the men, compared to the women. This is expected since Birka is a place for trading.

A comparison between the Viking Age places in Sweden and those in, for example, Denmark show that the variation in strontium values at Galgedil on Funen or Trelleborg on Zealand was not as large as at some of the places in Sweden. At Grødby on Bornholm, however, the variation in strontium is large. To some extent the reason may be that the geological conditions in Denmark are more homogeneous. Regarding Bornholm, the geological conditions are different which means that the individuals at Grødby may come from different parts of the island. However, compared to Birka or the island of Öland, the difference is evident. People that visited or settled in those places came from a large geographical area. It also seems that several of the places in Sweden had a proportion of non-locals (Fig. 45a, c and Fig. 50).

Have we been able to answer the question of which people were local to the place and which people were non local? With the knowledge we have at present it looks as if we have a rough picture of how mobile people were, and we can also discern certain patterns as regards the origin of the immigrants.

Someone knew how the deceased wanted to be buried

There are examples from the Viking Age of people being buried in their clothes, with all the brooches and accessories such as ornaments. In addition, various objects were deposited with the deceased, things that could be important to take along on the journey to the other side, to mark the individual's status, or perhaps even occupation. Not everyone belonged to the social group that was able to adorn their clothes with fine brooches. Those who were worse off could perhaps not afford to let their brooch be buried with them instead of being inherited. If one came from the locality it could be hoped that someone, if not always a relative, would look after the body, but what happened when foreigners died, for example a merchant on his travels? Did people always travel in groups, and were the travelling companions close enough to know what the deceased would have wanted in the grave? The other question is how long people retained their own customs and how permissive the local people were about these customs being practised. The answer to this, of course, cannot only be found through the objects deposited with the body, but also through demonstrable ceremonies. What is the link to the individual in such cases? Was the object received as a gift, did the person in whose grave it was found bring it back from travels abroad? Was the person originally from the place where the object came from? No doubt all three variants are represented in a cemetery, but let us see if there are any cases when both the individual and the objects show non-local traits.

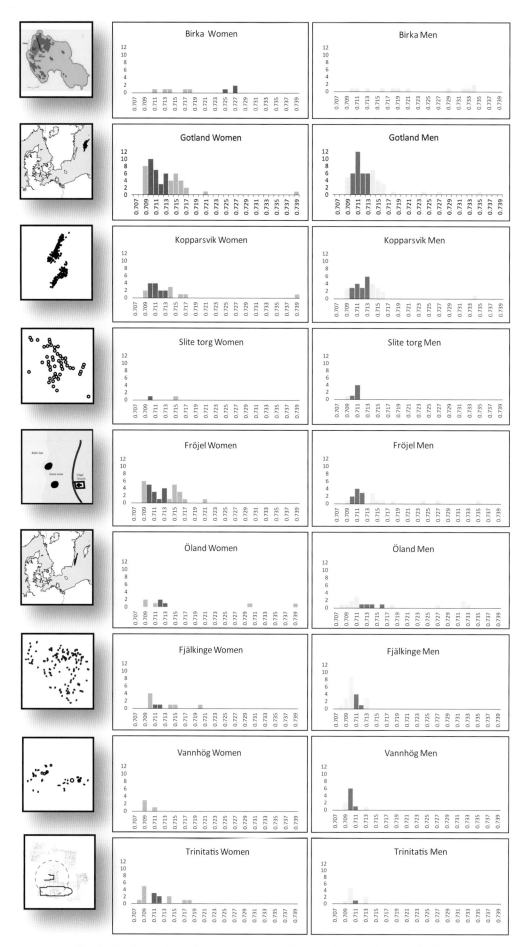

Fig 49 The figure shows the strontium values for men (blue) and women (red).

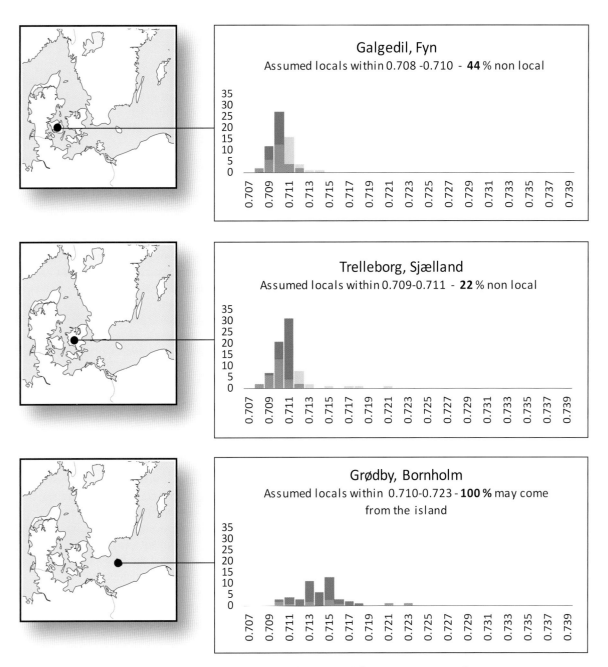

Fig 50 The results of strontium analysis from three different sites in Denmark.

In the Christian cemetery of Trinitatis the bodies, with few exceptions, have no grave goods. One can see, however, that certain funeral customs, such as rods in the grave, occur especially in the case of bodies with non-local strontium values. The use of grave rods may originally have been a German custom.[32] In the part of the cemetery where most of the non-locals are buried, that is, on the edges, there is also variation in the types of coffins, for example, wicker coffins, coffins made from old troughs, five coffins made from boat boards with remains of sphagnum between the boards. The boats are thought to have been built on the south coast of the Baltic Sea, in other words, in Slavic lands.[33] This is also the part of the cemetery where we find individuals with low strontium values, 0.708–0.709. To be able to show that it was non-local individuals whose wishes were respected in the form of the grave goods deposited with them, it is not enough that the grave is exceptional in character; it must also differ noticeably from the others.

Perhaps we have two such cases at Kopparsvik, where both finds and strontium values indicate that we are dealing with the burial customs of non-local individuals. One case is the individual in grave 230, a grave located

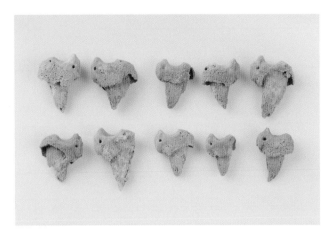

Fig 51 Ten bear claws found in the male grave 230 at Kopparsvik. (Photo Jonny Karlssson, Swedish History Museum, Stockholm)

in the southern part of the burial ground. The grave was dug into humus-mixed shore gravel about 40 cm below the present ground level, and the skeleton was covered with a rectangular stone packing. The grave contained a man who died at the age of only 20–25. He was a relatively tall individual, about 176 cm, and he was placed stretched out on his back with one leg straight, the other slightly bent, and the feet together. Over the skeleton and in the stone packing were remains of wood which may have been some form of wooden lid. In the grave were found a total of eight bear claws, five of them a short distance from the head, one inside the elbow, and one at each foot (Fig. 51). Bears do not occur naturally on the island, so a bearskin must have been brought there.

A survey conducted by Petré and Wigardt[34] shows that bear claws have previously been found at 23 cemeteries in 37 graves at various places on the island. Thus, this is the 38th example. Most bear claws are found in cremation graves from the Vendel Period, 49% (18/47), and previously only two such graves have been found from the Viking Age, both cremation graves. Finds of bear claws occur in both men's and women's graves. The claws have been viewed as signs of both courage and strength and, like teeth, they were used as amulets or talismans.[35] Finds of bear claws in graves have also been interpreted as reflections of high status and have been discovered in several large mounds (Petré 1980). Their occurrence also reflects the regional fur trade and can even indicate considerable long-distance trade.[36] Should the find of the claws with the man in Kopparsvik be interpreted chiefly as a sign that he was buried in a bearskin? What is there to indicate that? Is the man a person of high status, is he from the island? Apart from the bearskin, two ring brooches, a belt buckle, a strap-end mount, and an arm ring, all of bronze, and an ordinary knife, a key, and a long one-edged knife with bronze mounts on the sheath show that this was not just anybody. The knife, a *scramasax*, was a weapon rather than a tool in the Viking Age.

What does the strontium analysis tell us? The result (0.714) shows that he was not originally from Gotland, nor was he from either northern Sweden or the Mälaren valley. Perhaps he did not hunt himself, but was a merchant who traded in expensive goods. He needed the *scramasax* to be able to defend himself in the event of conflict.

The other highly distinctive grave is that of a woman, grave 112, located in the northern part of the burial ground at Kopparsvik. Like many others, she was dressed for the burial and a part of her dress was an animal-head brooch. What is special about the grave, however, is that no fewer than 14 Arabian silver coins were strewn over much of her body when she was buried; they lay from the upper arm down over the hip and both thigh bones. There was one Abbasid coin and 13 Samanid coins. The Samanids came from a Persian dynasty who had liberated themselves from the Arabian caliphate and subjugated areas both east and south of the Caspian Sea. The coins are dated to AD 894–936. Nothing like this has been found in any other grave at Kopparsvik. On Gotland as a whole, coins have been discovered in a total of 43 graves; in a large share of the cases, 58%, it is just one or two coins. In the other contexts the number of coins is between 3 and 14, and only one grave has more than this, 44 coins.[37] It was thus not a common act to deposit so many coins in a grave.

The woman, who was over 60 years old when she died, had incurred a fracture of the lower arm; although this had healed, it had a secondary effect of arthrosis in the wrist and had also led to deformation of the wrist bones. Who was she, where did she come from, and what is the meaning of the combination of animal-head brooches, which are considered to be so typically Gotlandic,[38] and the Arabian coins? Does it mean anything at all? Perhaps she was a wife, a beloved grandmother, whose husband or son was a merchant and had plenty of coins hidden away. What is interesting in this context, however, is that the woman is not originally from the island; she was not Gotlandic.

Four possible reasons are suggested for why was she was given so many coins with a strontium value of 0.717 in the grave. When it comes to coins in graves there are the following hypotheses. One is that they were brought along as wealth to the other side; the second is that they were payment for the journey to the realm of the dead (coins for Charon); other suggestions are that this was a religious symbol, a decorative element, or a protection against evil spirits from occupying the dead body.[39] Since there are 14 coins, they cannot be Charon's obol (since it is not one coin and it was not placed under the tongue). Nor are they coins with crosses which could indicate a Christian symbol, or decorative elements since they are not perforated to be worn as ornaments or amulets. It should be mentioned that she is not the only non-local woman who was buried with the typical Gotlandic animal-head brooches. Since the coins are scattered over the body, it is possible that those who arranged the burial poured the coins over her. It is less likely that she had them in a purse and that animals spread the coins over the body. What can the scattered coins symbolize? Are they just a sign of wealth? Since so few people were given such grave goods, one may wonder whether it might be a custom brought from elsewhere. One way to answer these questions is, of course, through strontium analyses of other similar finds to see whether there is a connection.

At the burial ground in Slite there was an example of the opposite, that is, the object in the grave suggests an import but the strontium value indicates that the person is local. This is the man in grave 8, who was found with a buckle for a money bag in the grave. The bag may have been made in Birka but with eastern inspiration.[40] The strontium value (0.711) suggests that he may very well have been a Gotlander. In other words, the man could have acquired the bag on his travels, or received it as a gift or exchanged it for something from a merchant visiting the island.

As for Birka, it may be reiterated that only 19 of the 37 excavated inhumation graves allow sure identification of the skeleton and grave number. In one of the 19 cases there is also an interesting link between finds in the grave and the strontium value. This is grave 585, which contained the skeleton of a woman who had been buried in a chamber grave. On the skeleton were two costume brooches, in this case oval brooches, with traces of gilding; a white bead lay under one of the brooches. Also in the grave was a pot, an oak disk that had been the lid of a box. The box has been compared with a similar find from Ketting, on the Danish island of Als.[41] A closer study of the woman's strontium value (0.711) shows that she is not from the Mälaren region but either from Gotland, Öland, Skåne, or Sjælland. There is thus a possibility that the find accompanied the woman on her journey northwards. At the cemeteries in Vannhög and Fjälkinge there are no similar cases where we have been able to associate finds from foreign parts with non-local strontium values. Have the strontium analyses brought us closer to an answer to the question of whether non-local people were buried in the local cemetery or not? The problem with that question is that we do not know whether the outsider was visiting the place or had lived there for some time. In other words, there is nothing to suggest that different customs were applied to non-locals.

Did everyone come here voluntarily?

Migration or mobility during the Viking Age was not always voluntary, it could involve abduction. The most frequently quoted contemporary eyewitness account of how captives could be treated comes from the Arabian chronicler ibn Fadlan, who describes how a slave woman was sacrificed for her master. Attempts to trace those captured in war and to make them visible have long been made by historians,[42] but in the last twenty years archaeologists have also searched for criteria to identify this group.[43] Among other things, it is the occurrence of "double graves" that has led to speculation about the presence of this social group.[44] Various isotope and aDNA studies have also been tried as a way to demonstrate the presence of slaves.[45]

Being a slave, however, was not a static position. A person could be taken into slavery as a child, be born as a slave, be enslaved as an adult, or even voluntarily choose slavery.[46] A person who had become a slave could also be liberated, but the procedure for this could take a long time, sometimes up to four generations.[47] There were also differences in the status of slaves, for example between one who was bought as a slave and one who was born into slavery, known as a *fóstri*. The latter category was more trusted and could be allowed to carry keys and own property.[48] The fact that slavery could take such different forms naturally makes it even more difficult to detect evidence for slaves, that is, to distinguish them from other people in archaeological grave material.

As mentioned in the previous chapter, the position of the dead in the grave could vary, but the most remarkable position is face down. Zachrisson, in an article about slaves, has discussed that those who were buried face down at Kopparsvik could be individuals who were subjected to various kinds of punishment.[49] Zachrisson also refers to an essay by Funegård Viberg,[50] who notes that a number of the dead who were placed on their stomachs lack body parts, such as the head. A systematic study of descriptions of the burials at Kopparsvik show however, that there is no difference between those buried face down (14/38) and those buried on the back (40/145) with respect to lacking skeletal elements. The difference is not statistically significant (p = 0.3181). In several of the descriptions, however, we see that the graves have been damaged by later burials, which is probably the main reason for the missing parts. If it is the case that those buried face down were slaves, it might be envisaged that they came from elsewhere to a greater extent than those who were laid on their backs and thus ought to have different strontium values from the locals. Of course, there is the possibility that that they are second- or third-generation slaves and therefore were born in the place. The results show that the man in Vannhög was from the locality whereas the man in Fjälkinge was one of those with low strontium values (0.709) who came from outside. At Kopparsvik, where the largest accumulation of this practice was found, it happens that eight of the 42 underwent strontium analysis. Of these eight, five (62%) individuals had strontium values indicating that they were not local. Among those placed on their back or side, 40 have undergone strontium analysis, showing that 18 (45%) were not local. At the cemetery in Fröjel, the two who were buried on their stomachs were not of local origin, but many of those buried on their backs were likewise not from the locality. There is thus nothing to suggest that those who were buried on their stomachs were more often non-locals than those buried on their backs.

4

Health and care for the frail

It is a challenge to paint a picture of health status, living conditions, and the care of the sick in prehistoric times. This is particularly difficult since the term "health" can include both physical and mental well-being, and also because it can be perceived differently by people.[1] What is it that affects health, and what do analyses of health say about living conditions? Factors that affect health are heredity, sex, and age, but also the social environment in which we grow up, our housing, the availability of food, sanitation, whether one has the opportunity to grow up with both parents or at least with other close adult relatives, or brought up by strangers or forced to fend for oneself. What were conditions like in Viking Age society? The general picture we can paint based on the grave material is of course limited, but it still gives some insight into the living conditions of the time. Diseases such as chronic infections and injuries are affected by the living environment, hygienic conditions, dietary habits, the availability of food, as well as accidents and violence. Can we detect signs of care for the injured, the handicapped, and the sick? Was there acceptance for people who differed from the norm?

Analyses of skeletal material from different parts of our history show that chronic states which can be detected in the skeleton account for around a third of all the diseases we can trace. The diseases in question are above all those with a chronic course, which meant that people survived for a long time. To simplify, we can therefore say that, if we find that many people in a population died young, they had a poorer state of health than a group with a higher mean age who show pathological changes in the skeleton. One could say that the best physical health is seen in those who had no signs of skeletal changes and lived to an old age.

Trauma in the form of fractures that started to heal or were fully healed can also be detected. The three criteria used in today's society to compare health over the whole world are mean longevity, infant mortality, and stature. The first two are difficult to study in mortuary material, partly because we seldom or never find all those who were buried, but also because our present methods for determining age do not allow us to state the exact age of those who grew old. The third parameter, stature, is easier to ascertain, and here it is also possible to make comparisons between different groups.

Earlier studies of health in Norse society in the Viking Age concern both individual diseases and give a general picture of health in particular places. We can read about this in reports, in larger amalgamated studies and dissertations.[2] A large share of the grave material from the Viking Age, however, is represented by cremations, where the degree of fragmentation does not contribute much to our knowledge of health matters.

"Tall as palm trees"

With an average stature of just over 172 cm, men in the Viking Age were taller than many of those they met on their travels in Europe.[3] Some men from Scandinavia were very tall, up to 2 m, and could therefore be described in ibn Fadlan's words, "tall as palm trees".

Stature, which is regarded as a health parameter, has varied throughout the history of humanity; it is not the case that we have continuously grown taller in the course of history, since there have been periods when people have been both taller and shorter.[4] Osteological studies show, for example, that people in the mid-19th century were as short as in the Stone Age, while people in the Roman Iron Age (AD 0–400) were as tall as today. Variations in stature are not specific for Scandinavians, however; they can be seen

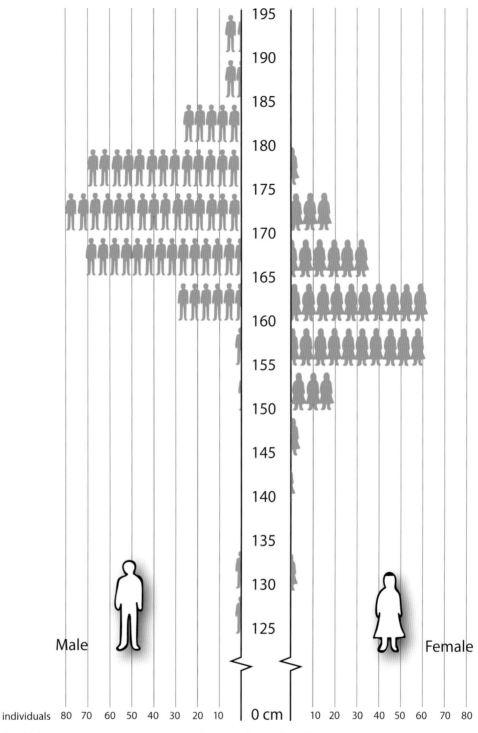

Fig 52 Variation in stature among men and women during the Viking Age. Based on 595 individuals.

throughout history all over the globe, changes which can only be studied over a long time by examining skeletal material.[5] From the middle of the 19th century until today, the curve for stature has risen sharply for several countries in Europe.[6]

Stature in skeletal material is usually calculated on the basis of the greatest length of the femur, as is also done in this study. Altogether we know the stature of no fewer than 595 adult individuals from seven of the burial grounds. For women in the Viking Age we see a variation of 30 cm between the shortest (145 cm) and the tallest (175 cm), with a mean of 160 cm, varying somewhat depending on where people lived. In men the variation was almost 40 cm, from 154 cm to 193 cm and with a mean of 172 (Fig. 52).

Although stature varies somewhat between the different places, the difference is not statistically significant. The stature of Viking Age people in general indicates that they were as tall as in the Middle Ages. We can see, however, that the really tall men and women lived on Gotland. There are several reasons for the variation in stature, and it is believed that both heredity and living conditions are significant. Above all, the circumstances of the mother during pregnancy are important, but also the conditions for the child.[7] Social group affiliation, the ability to provide for oneself, family size, crowded housing and sanitary conditions which in turn affected nutritional status, and exposure to infections are relevant factors.[8] Although there are other factors that affect stature, it is thought that it can be used generally as a measure of living conditions.[9] This means that the conclusions we can draw about health and living conditions in the Viking Age suggest that they were fairly good, equivalent to conditions in Scandinavia at the start of the 20th century.

Toothless or shining white?

Although people of the Viking Age are best known as seafarers, the majority lived by agriculture and animal husbandry. They grew barley, wheat, oats, rye, and peas[10] from which they made bread, porridge, and soup. They used malted barley to brew mead which was sweetened with honey. Osteological analyses of animal bones show that they kept cattle, pigs, sheep, and hens[11] which provided the people with meat, milk, and eggs. Although honey was the only source of sugar, the starchy food was sufficient to give people caries. Because knives and teeth were the only available implements for cutting food into smaller pieces, and because the stones of the hand-powered querns were coarse and gave off small stone particles, there was extensive wear on the chewing surfaces of teeth. Large grains of sand in food could also cause small fractures of the tooth,[12] and remains of food could easily stick in the small cavities and cause caries. The osteologist Pia Bennike has shown, however, that toothpicks were used, sometimes so much that they left marks on the teeth.[13] Dental health was particularly poor in old people. Among those buried at Fjälkinge this is very clear; the average number of remaining teeth was about 30 for those aged 20–40, while those aged 60 had fewer than half left (Fig. 53). It was above all molars that were lost. Caries or fractures on the teeth had also occasionally led to inflammation in the jaws (Fig. 54). Among the older people at Kopparsvik and in Lund the pattern of dental health looks much the same. Many of those who died young, however, had very fine and intact teeth (Fig. 55).

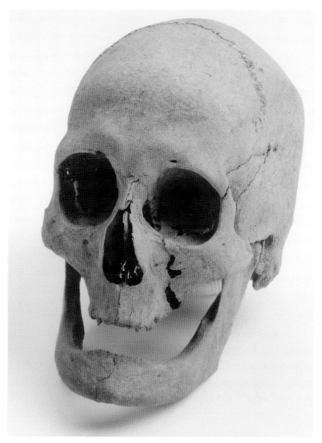

Fig 53 An old woman, above 60 years of age, who was nearly toothless. (Photo Historical Museum, Lund)

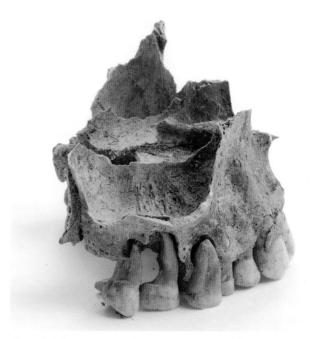

Fig 54 Inflammation in the upper jaw caused by bacteria that spread into the jawbone, probably due to a caries infection. (Photo Historical Museum, Lund)

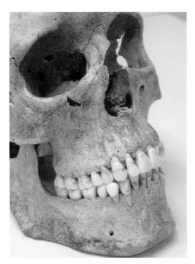

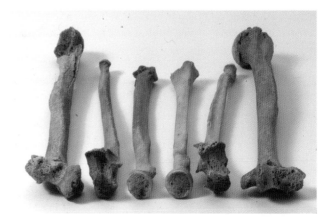

Fig 56 Elbow joints with lesions in form of small perforations, most probably due to rheumatoid arthritis. (Lars Westrup, Kulturen)

Fig 55 Healthy teeth in a young adult individual from Kopparsvik. (Photo Caroline Arcini, The Archaeologist, National Historical Museum)

Far-reaching caries attacks could cause intensive pain as well as protracted toothache, which raises the question of what kind of help was available. The tooth could be pulled, of course, but many skeletons show that this does not appear to have been a common measure. Although lack of teeth can be a sign that they were extracted, the evidence against this is that there are far more examples of teeth that are totally corroded by caries. Occasionally all that can be seen is corroded roots. Of course this may be because attempts were made to extract the decayed tooth but only the crown came away, or else the caries attack continued, and the crown was entirely corroded.

What form of painkillers were available at this time? Finds of pollen and seeds from archaeological excavations in recent years have shown that there were medicinal plants in Scandinavia even before the coming of the monasteries. There was extensive contact with other parts of Europe, and the Vikings probably brought home seeds, cuttings, and grafts to introduce to Scandinavia. Among the finds of plants that could be used to heal and relieve pain we find opium poppy[14] and henbane.[15] In Henrik Harpestreng's herbal from the 13th century we read, for example, that henbane kills the worm that causes toothache.[16] The level of alcohol in mead was presumably not high enough to anaesthetize, and if it was sweet from the honey it could have made the pain worse.

Joint problems

Joint problems of various kinds have followed humans from the very oldest times.[17] By far the most common joint lesion is the one that goes under the name "arthrosis", which means that the cartilage protecting the joint from damage by knocks is cracked or abraded. "Wear" is the designation used of this ailment by people without medical schooling. In a way this is a correct term in many of the cases, but the protective cartilage can also decay in connection with injuries or be corroded by disease. X-ray examinations of persons living today show that arthrosis is more common in the ageing part of the population, but can also arise early if the joint is damaged in early years or if the person was born with a malpositioned joint. If the joint is affected by disease, as can happen both in childhood and in adolescence, the cartilage also undergoes strain and arthrosis can arise at younger ages.

Studies of skeletal material from both the Viking Age and the Early Middle Ages show a high frequency of arthrosis. In most cases we see it in skeletons judged to be middle-aged or older, but in certain cases it is ascribed to underlying diseases, both congenital and arising later in life. Several of the afflicted individuals have not only the ordinary age criteria but also decalcified skeletons and heavy tooth loss in the jaws. Just as in the large assemblages of grave material from Lund and Kopparsvik, there are also examples of individuals with diseases that caused malposition at an early stage, and fractures that put the joint out of position while the injury was healing and therefore led to arthrosis. At the Skämsta cemetery, which will be described in more detail later in this chapter, there is also one case where a congenital disease is the underlying cause.

Rheumatoid arthritis is another disease that is rare today and also in prehistory. From the Viking Age, however, one suspected case has been noted. It is a woman who died at the age of 50–60. Her hands, elbows, knees, and ankles show lesions in the form of large numbers of small cavities, suggesting that rheumatoid arthritis was the cause (Fig. 56).[18]

Injuries in the form of fractures which result in malposition when they heal, or if the bone comes out of joint, also occur, albeit not so commonly. One such example is a woman in Fjälkinge whose shoulder dislocated and never returned to its original position.

Less common joint problems are diseases such as Fong's syndrome or Nail-Patella syndrome as it is also called. The word syndrome means that a set of symptoms occur simultaneously. This is a disease that arises either through a spontaneous mutation (20% of cases) or through inheritance. In the latter case it affect one in 50,000 births. The skeleton that has been noted with this lesion was found at Fjälkinge, and it belonged to a woman who was about 40 years old when she died. She may have been afflicted by the disease through either mutation or inheritance. The syndrome involves disturbances to renal function and shrunken and underdeveloped fingernails and kneecaps.[19] A thousand years later we can identify the disease through the underdeveloped kneecaps but also as growths on the back of the pelvis, known as iliac horns, which are typical skeletal lesions (Fig. 57). The reason Fong's syndrome is mentioned in the chapter on joint disorders is that underdeveloped kneecaps cause joint problems with the consequence that the individual has a waddling gait and the kneecaps easily get out of joint.[20]

Lesions in the small and large joints of the back are among the most common joint ailments. Most common is degeneration of the discs between the vertebrae, and the stabilization of the back to compensate for this results in the formation of new bone on the vertebrae, causing spondylosis. The cartilage protecting the small joints in the vertebrae can also be damaged, as can the cartilage in large joints, with the same consequence, that bone wears against bone and eburnation (an ivory-like surface) arises.

Although the back is less mobile than the extremities, it is full of joints large and small, and it is subject to disease and especially constant strain. At the same time, we humans really need our joints to function adequately, not least in our day-to-day tasks. The most common ailments in the back are those that arise when lesions occur to the shock-absorbing discs between the vertebral bodies. These lesions are seen above all in middle-aged and older individuals.[21] But the back can also suffer malformations which cause problems in the early years of childhood. At Vannhög there were two women who had incurred congenital lesions to the cervical spine. This disorder, known as Klippel-Feil syndrome, would have made it difficult to hold the head looking forward. The problem is due to a deformation that occurs between the 3rd and the 7th week of pregnancy. It could thus be seen at birth that the child was different.

Today this deformity is present in one child in 40,000.[22] The disease is slightly less common in girls than boys. In most cases it occurs sporadically through mutation, but in 2–3% of cases the cause is hereditary. If the gene is dominant there is a 50% risk that the child will be affected; if it is a recessive gene the risk of passing it on falls to 25%. One of the woman died young, at the age of 19–20, while the other reached an age of 50–60. The young woman had four of the cervical vertebra fused together in one block, causing

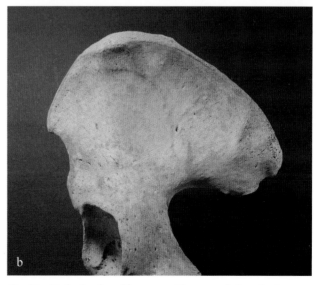

Fig 57a Underdeveloped kneecaps. The normal size of a kneecap of an adult individual to the left in the photo. 57b Top: Right pelvic bone with a bone growth protruding from the ilium (a iliaca horn), bottom: A normal bone. (Photo Staffan Hyll)

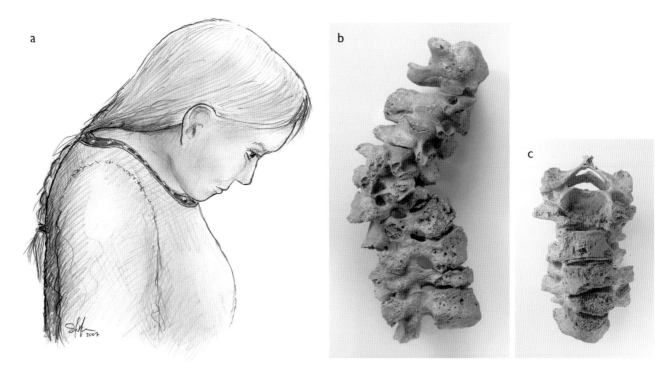

Fig 58 Four cervical vertebrae fused into a block, which resulted in the head of this young female leaning forward, with a reconstruction of her appearance in life. (Photo and illustration Staffan Hyll)

her to bend forward; she was forced to look at the ground for most of her short life (Fig. 58a and b).

The older woman had a milder variant with only two vertebrae fused. Apart from the lesions in the neck, the disease can also result in deformities in other organs, such as heart defects, malformations in the brainstem or the spinal cord, which in turn can affect the nerve cells that control breathing. In view of the size of the communities at this time, this must have been a rare sight. Of the two women at Vannhög with this syndrome, the older woman's problem may not have been so noticeable. Since there is a certain hereditary risk of being born with this syndrome, it is not impossible that the two women were related.

Everyday accidents and battle traumas

The designation Viking is intrinsically associated with violence. Every publication, lecture, film, and television programme about the Viking Age considers the Vikings in different forms of conflict. The authors of the book *Vikingar i krig* ("Vikings at War") describe Viking Age society as militarized, with everyone being involved in some way, even the women.[23] Their role was to prepare their sons for the demands of society, and to preserve the memory of the fallen, besides which they are regarded as intriguers urging their menfolk to fight. A contemporary Arabian source also gives insight into how sons were brought up to rely solely on their weapons, and it depicts a society pervaded with threats of violence and death. Does this picture match reality? A study of the skeletons shows that this is scarcely the case.

That the Viking Age was an age of conflict is thought to have many explanations. It was a hierarchical society, with kings, jarls, and chieftains at the top, under them the free farmers with their own land, and at the bottom the landless peasants and slaves (thralls). The free farmers could lease land to people without land or run their farms with the help of slaves. The farmer had to protect the people on his farm. At the start of the Viking Age the political and economic situation in Western Europe is deemed to have been stable, with vigorous economic growth. Trade in luxury goods was extensive. The Norse commodities consisted of amber, furs, walrus ivory, leather, honey, and down. Trading, however, required control, and those who exercised this control could amass wealth, one way being through profitable piracy, a kind of Viking raid. At first the plundering expeditions were mounted between the Norse lands but they were later extended to France and England. It used to be believed that population growth forced the Norse to emigrate, but today it is thought that the growth was not sufficiently large to trigger emigration. On the other hand, the fact that only one son could take over the farm had the effect that those who did not inherit set off in search of honour and respect elsewhere. There could also be conflicts between heirs, sometimes with a bloody outcome.

What form did combat take in the Viking Age, and what type of protection did the fighters have? Our knowledge of this comes from written sources, from picture stones, runic inscriptions, the famous Bayeux Tapestry, and not least of all from archaeological finds. In this connection the Gjermundbu find from Norway is by far the best preserved. The grave finds here show us the swords, spearheads, axes, shield bosses, and chain mail of the time, and the only known surviving Viking Age helmet.[24] It is also thanks to archaeological excavations that warships of various sizes can be studied, for example, the Skuldelev ships at Roskilde in Denmark, or the smaller ships recently found on Saaremaa.[25]

Weapon training was important during the Viking Age and was carried on from early youth. A knowledge of weapons and good physique were essential, requiring practice in wrestling, swimming, riding, and hunting. All free men were expected to bear arms, and they rode to the assembly, the *thing*, with complete equipment or else they could be fined. According to the laws of both Gutating and Frostathing, men had to carry a spear, shield, sword, or axe.[26]

How many set off on military expeditions is not known. As is often the case in the written sources about war, there is a tendency to exaggerate in order to show that the threat was greater than it was, and perhaps also to make the story more exciting. It is, of course, the victors who write history. The Vikings used the technique of surprise. They lowered the masts on their narrow, shallow boats, which could moor anywhere. They attacked quickly, shouting and roaring to paralyze the enemies who were unprepared and did not have time to assemble in defence.[27] They had previously reconnoitred the place they were going to attack, and the assault preferably took place early in the morning. Their booty consisted of captives and valuables. When seizing prisoners for the slave markets they favoured young men and women, rejecting old people and children, the lame and the infirm. Nor did they capture adult men, who could be a security risk. The Vikings stole anything of value and then burned the houses. If they encountered hard resistance they preferred to disperse rather than fight. They displayed personal bravery and strength rather than moving forward in a group, thus forcing the enemy to split up their forces. The raids took place during spring and autumn.[28]

When the Norse fought in a larger army for a king, they likewise used stratagems, such as not to show all their forces at the same time and thus make the enemy believe that their numbers were smaller. Another tactic of the Vikings was to spread out and attack from the side or from behind. Often there was no clash. When they fought at sea it was not in open waters but in fjords and bays, or close to land. The use of horses did not begin until near the end of the Viking Age, chiefly by the elite.[29]

What are the physical traces of the conflicts in the skeletal material from the period? To what extent are there injuries from weapons? The burial place with the largest occurrence of weapon graves in Sweden, where we might also expect injuries to the skeletons, is Birka. Unfortunately, the skeletal material is very poor and thus not of much assistance. Moreover, several of the weapon graves are cremation graves, which do not yield any information about battle injuries, whether healed or not. At the other cemeteries studied here, weapons in graves have only been found in significant numbers at Kopparsvik and Slite. Yet those who were not buried with weapons could also have been in battle and show marks of this on the skeleton.

If we look at skull injuries caused by sharp weapons such as a sword, axe, or some form of war hammer, these occur in Lund during the period from the 990s to the start of the 12th century in just over 1% of the men and the women. It is impression fractures that dominate, but there are also lethal cutting injuries. The impression injuries arose through blows with a blunt instrument, with sufficient force to knock the person down without breaking the skull. Yet the occurrence of skull injuries is not particularly high compared with the subsequent 200-year period, the 12th century to the start of the 14th, when over 6% of the men and 2% of the women suffered in this way. In contrast, injuries to the bones of the extremities arising from a sword or axe were rare during both periods. Only one case has been noted in these materials. We may note that in the early period we see traumas resulting in fractures in different parts of the body in about 5% of the men and women, to be compared with almost 12% and 8% respectively in the next period from the 12th century to the start of the 14th.[30] Fractures may, however, have happened in connection with everyday accidents and need not have anything to do with fighting or conflicts.

At neither Vannhög nor Fjälkinge are there any injuries that can be associated with sharp weapons. Both sites, however, have individuals with healed fractures. At Vannhög there were three men and one woman with healed fractures. One of the men, who was buried in one of the mounds, had several injuries in the bones on the left side of the body. The shoulder blade was partly crushed, the clavicle was broken (Fig. 59), and the knuckle of the index finger showed the kind of injury one could expect to result from a fight. On the fibula there was a healed fracture up at the knee, and the toe bones appear to have been trodden on. The injuries may have arisen in battle, but could also be the result of a hard fight with a blunt object. Of the other men, one had a rib fracture, one a fracture on two thoracic vertebrae. The latter may have arisen if someone had hit him on the back with a blunt instrument, a cudgel. The woman has a fracture on a metacarpal.

At Fjälkinge there are seven individuals, almost 14% of the adults, who display traces of trauma, but none caused by sharp weapons. Two of the women incurred luxation in the shoulder joint. In one case the shoulder was dislocated backwards at an early age, probably in childhood, and it had not been successfully put back in place (Fig. 60). The result was that the joint could not be used. In the other case of

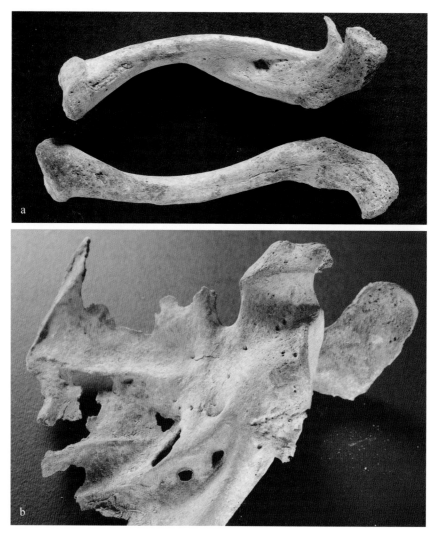

Fig 59a–b Shoulder blade partly crushed and associated clavicle with a healed fracture. (Photo Staffan Hyll, Riksantikvareämbetet UV Syd)

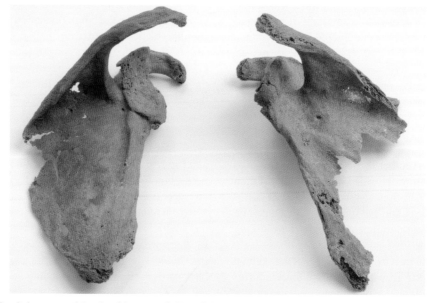

Fig 60 Luxation and a dislocation of the shoulder joint (left) and the unaffected (right). Fjälkinge cemetery. (Photo Kristianstad Museum)

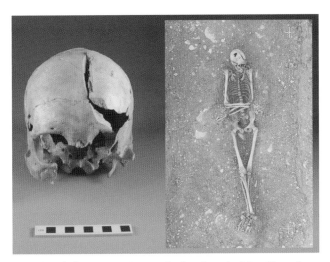

Fig 61 Lethal weapon injury to the head, an incision 11 cm long, Kopparsvik cemetery. (Photo Jonny Karlsson, National Historical Museum)

dislocation, the woman's shoulder ended up on the front of the body. This may have arisen due to a fall on an outstretched arm. This woman also suffered a fracture on her lower left arm, which could possibly have happened on the same occasion. The third woman was old, and her skeleton was badly decalcified; she had probably incurred her fractures in an accident, breaks on the lower part of the upper arm and the front of the pelvis. The fourth woman had lost one thumb, either through an accident or an amputation. The men had fractures of the ribs, the lower arm, and the collarbone.

At the large and almost entirely adult-dominated cemetery of Kopparsvik on Gotland, lethal weapon injuries have been noted on just two individuals. One man had an incision 11 cm long on the head (grave 154) (Fig. 61) and another has slashing injuries on the right side of the mandible, the right femur, and the tibia (grave 4). Both were tall individuals, 185 cm. Neither of them had weapons in the grave, only a knife each. Both had injuries that killed them. The man in grave 154 appears to have been in a fight or suffered an accident before he incurred a mortal blow in combat. He had previously incurred fractures of the ribs and the collar bone. Moreover, it looks as if he had some form of periosteal inflammation on both shinbones. The loss of a left front tooth before death was probably also due to a fight.

Of the total 242 men, 176 were buried in the northern part of the burial ground and the two men with weapon injuries are found here. Also observed in this area are a number of individuals with fractures on different parts of the skeleton, and yet another individual with the front teeth knocked out. As regards fractures, however, the frequency is not as great among men as women, 9%. By way of comparison it may be mentioned that no man with a fracture has been noted in the southern part of the cemetery, but there are two women. Fractures occurred on the upper arm, the lower arm, the femur,

the ribs, the pelvis, and the collarbone, and in one case also on several bones of the hand. Regarding trauma, individuals in all ages are represented but a large share of those whose age has been determined are young. At Kopparsvik seven individuals were found with weapons in the graves, but none of them displays any weapon injuries or fractures.

Among those buried at Slite, minor skull injuries can be seen on two individuals, both men, and nine individuals (just over 30%) have healed fractures on the bones of the lower arm, the back, the lower leg, and the ribs. All but one are men. The two with skull injuries were buried with weapons, one with axes and spears and the other with a *scramasax*. The man with the healed fracture on the fibula was buried with his weapons, an axe, a spearhead, and iron fragment which is thought to be from chain mail.

Even though the skeletal material from Birka is poorly preserved, one individual displays injuries caused by sharp weapons. This is the man in grave 138, who incurred blows to the left side of the skull and the upper arm. He was not buried with any weapons, however.

Traces of weapon injuries were thus fairly rare in a time that is known for its violence. What can the reason for this be? Did the surprise tactics of the Vikings mean that in most cases they managed to escape without injury, or was their equipment more efficient than we know? Is it possible that the men who actually were injured did not survive and were buried abroad, or that they died of their wounds on the way home and were buried at sea? Some, of course, could have incurred injuries that did not leave any marks on the skeleton. There is presumably a hidden figure that we have no way of knowing, especially as regards the cremation graves. Perhaps it is also the case that some battle injuries are concealed among the fractures, since the men also fought with blunt weapons.

If the warriors died in foreign countries, they should have ended up in graves outside Scandinavia. Just a few years ago, two mass graves from the Viking Age were discovered, one with executed men, the other with injured men. The former is to the east at Salme on the island of Saaremaa (present-day Estonia) and can be dated to the time around AD 750, that is, the start of the Viking Age. It consisted of two boat graves, one with seven skeletons, the other with 28. The boats had been drawn up about 100 m from the water and were partly covered with stones and soil. In one of the boats none of the men showed any weapon injuries, while several men in the other boat had chop marks on different parts of the body. In the latter boat almost all the men were buried with their weapons.[31] The interpretation is that the burial of the crew on the boat with the weapons had taken place in haste. That the men in the other boat had not been buried with equipment may indicate that they were of a different social status. Strontium analyses suggest that the men in both boats may have been from the mainland of eastern Sweden, more specifically the Mälaren region.[32]

The other mass grave was located in Dorset in southern England. It is the sad reflection of a mass execution. From strontium analysis it was judged that the men, who lived in the period 900–1000, came from Scandinavia. Their average age was rather low, most of them being young men between 18 and 25, but there were also a few teenagers. The oldest individual in the group was over 50. What they were doing there and why they were executed is unknown. No finds were buried with them; brooches on clothes and weapons had been removed before they were covered.

There are several suggestions as to what event this was. The dead, for example, may have belonged to the band that pillaged the island of Portland in 982 or may have taken part in one of the Viking attacks on Dorset in 998, 1015, or 1016. They may have been executed by other Vikings fighting for the English, or they could have been Scandinavian settlers who were executed in connection with Aethelred's purge of Danes in 1002. Several of them had traces of healed fractures on different parts of the skeleton. Injuries to the foot were most common, followed by rib fractures.[33] According to scholars, the number of fractures among the men from Dorset is much higher than in other contemporary material, resembling more the occurrence among the men who died in the battle of Towton in 1461. One of the men had one leg that was shorter than the other due to a fracture on the upper part of the femur. No earlier weapon injuries have been observed, however.

Apart from the complete skeletons, scattered bones were retrieved during excavation, among them bones with fractures and infections, for example, a man with extensive bone marrow inflammation. It must be said, however, that it is difficult to compare frequencies of fractures as a whole group since the preservation of bone can differ so much. It is fairer to compare the occurrence of fractures on separate bones and the number of individuals. Let us therefore look at a comparison based on the postcranial skeleton in material from the different sites (Fig. 62). It shows that ordinary fractures were much more frequent than weapon wounds, at least regarding wounds caused by axe or sword. If we compare frequencies of weapon wounds during the Viking Age with the circumstances during the Middle Ages AD 1100–1536 we find that there are still more ordinary fractures than weapon wounds. However the presence of weapons violence was greater during Middle Ages than in the Viking Age.[34]

The dwarf

In Norse mythology there are both dwarfs and giants. The dwarfs were small beings who were believed to live underground, afraid of daylight. They were skilled smiths who could forge remarkable objects for the gods, such as Thor's hammer and Odin's spear. Thor's hammer was called *Mjölnir* and it always hit the target when he threw it and it always returned to his hand. Dwarfs were also believed to guard treasure hoards.

Dwarfs, or persons of short stature, actually existed as physical people in the Viking Age. In this study we have found three individuals who were disproportionate in their bodies and significantly shorter than other people. They are two men and one woman who were born with diseases as a result of which they grew no taller than about 130 cm. One of the woman and one of the men were found at the Skämsta cemetery in Uppland, (Fig. 63) the other man at Kopparsvik on Gotland (Fig. 64).

Dwarfism is caused by various hereditary skeletal disorders (skeletal dysplasia). There are about 200 such different skeletal disorders. The individuals in Skämsta were born with spondyloepiphyseal dysplasia; although the body is proportionate in that the spinal vertebrae are lower than normal and that legs and arms are short too, the individual is abnormally short. The disease is either caused by a mutation or inherited from one of the parents, that is to say, it is autosomal dominant. In Sweden each year 1–2 children out of 100,000 are born with the condition, and in Sweden today there are 20–30 people living with the condition.[35] In most cases it is a result of mutation. The children born with the disorder are shorter than normal, 35 cm instead of around 50 cm. An adult can attain a stature of 80–140 cm. Already at birth one can see that the child looks different. The musculature is weak, and many die just after birth. For those who survive, motoric development is late, the child has a waddling gait, and the malpositioned joints show arthrosis earlier than normal, as in the case of the two dwarfs from Skämsta (Fig. 65).[36] The connective tissue of the eye can also be affected, which can lead to shortsightedness.

The dwarf from Kopparsvik is judged to have been short because of a different skeletal disorder.[37] Arms and legs are short and the thigh bones are curved, while the back is of normal length. This has been diagnosed as a skeletal lesion that matches one of the more common forms of dwarfism,

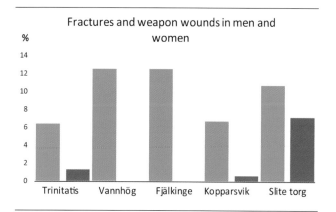

Fig 62 Frequency of fractures and weapon wounds in some of the investigated sites. Blue bars represent fractures and red bars weapon wounds both healed and unhealed.

Fig 63 Male dwarf from Skämsta cemetery in Uppland. (Photo Lars /Inge Larsson, RAÄUV Uppsala)

Fig 64 Male dwarf from Kopparsvik. (Photo ATA)

namely, achondroplasia. This occurs more often than the one affecting the people at Skämsta; today five children in 100,000 are born with achondroplasia.[38]

The man and the woman at Skämsta were buried in a farm cemetery. It has not been totally excavated; only six graves containing seven individuals have been excavated. The two dwarfs are buried in the same way as the other individuals in the cemetery, and they differ in no way from the other burials in the construction of the grave, the position of the body in the grave (both were buried on their backs), or the objects they were buried with. In the woman's grave there was a single comb and the man had been given a double comb and a knife. The double comb was made of elk antler.

The man and the woman in Skämsta had both incurred the same disease, and in view of the fact that it is so uncommon, the question immediately arose whether they were related. It is either father and daughter or mother and son, or else they are siblings. It is highly improbable, however, that there could have been mutation on two different occasions. The results of the strontium analysis show that they have different values, which means that they could not have grown up in the same place, It is of course possible that they were siblings if the parents moved when one of the children was about 4–5 years of age. If this were the case at least one of the parents must also have had the disease, in other words, was a dwarf. That individual, however, has not been found in the cemetery, but this has not been completely excavated; besides, graves have been destroyed by later development.

The other possibility is that they are father and daughter or mother and son. The skeletons were examined by Arcini in 1995[39] and now, after a wait of 20 years, the potential to answer this question will become a reality in the near future, since they are part of a major aDNA study of Viking Age skeletons from different parts of Northern Europe. Most striking, however, is the fact that they lived to become adults. This is also interesting in the light of the discussion among scholars about the exposure of children at this time. Two individuals whose deformity must have been noticed already at birth were obviously cared for rather than rejected. Moreover, this handicap cannot have been an impediment to marriage, since there are more than one of them and it is unlikely that they represent two separate mutations. In other words, this is a clear example showing that people in the Viking Age were inclusive rather than exclusive.

The dwarf at Kopparsvik on Gotland also seems to have been treated like other individuals, at least as regards the way he was looked after before burial. The two ring

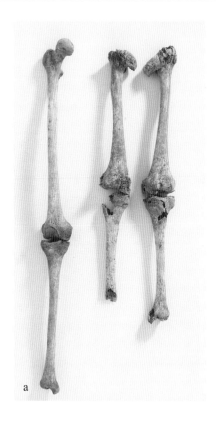
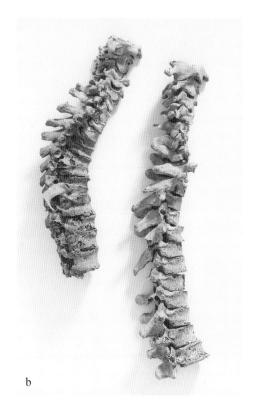
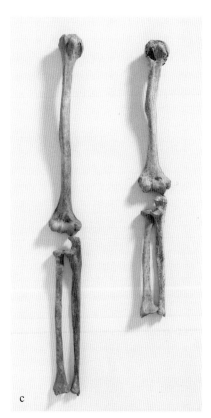

Fig 65a–d The short legs and deformed hip joint, flat vertebrae resulting in a short spine, short arms in the male dwarf and the short legs in the female dwarf, both from Skämsta. (Photo Staffan Hyll, Riksantikvarieämbetet UV Syd)

brooches show that, like the others in the same cemetery, he was buried in his clothes and the grave did not differ in form from the others.

The three dwarfs, with their unusual body proportions, had a distinctive appearance right from birth. When they become old enough to walk, people could note that they had a waddling gait and that they were not growing. At first their ability to move must have been fairly good, but they suffered from worn joints much earlier than others and were thus disabled. They may also have had other symptoms which are not observable on the skeletons.

Leprosy: noseless and numb

Leprosy, a disease that can disfigure a person beyond recognition (Fig. 66), has accompanied humanity for a long time. It could start with the eyelashes coming loose, the skin becoming hard in several places on the face, and large nodules arising. In certain cases, spots of varying size appeared on the skin, where the colour paled, and feeling was lost. The nerves in the peripheral parts of the body were affected, which led to paralyzed muscles and loss of feeling. As a consequence, the person could easily incur injuries, wounds, and infections. The skeleton was affected too, as the bones of the hands and feet, and the small piece of bone in the lower part of the nasal bone, regressed. For a person who had lived with the disease for a long time, the hands and feet could look like lumps with nails because the fingers had regressed (Fig. 67).[40]

Leprosy is an infectious disease which gained a strong foothold in Scandinavia at the end of the Viking Age. The oldest case of leprosy in Scandinavia has been found in a cemetery in Halland and dated to AD 70–570 (Fig. 68).[41] It was a woman and she was buried in a mound. In the mound the archaeologist found three burials, in the bottom there was a cremation, then the women was buried, and on the top of her another cremation was found. The palaeopathological signs like porosity of the nasal spine and resorption of the nasal spine and the alveolus to the front teeth of the upper jaw are indicative of leprosy. In several burials from the late Viking Age leprosy appears once again, sometimes not just individual cases but whole groups. It is difficult to say anything about the prevalence of the disease in the intervening period 500–900 because most of the grave material from this period is cremated.

Before the Viking Age leprosy was well known on the continent. In the 5th century several leprosoria were opened in France, places to which lepers were moved to be isolated. Osteological studies of grave material have reported cases of leprosy from Neuville-sur-Escaut in northern France and from Vaison-la-Romaine in the south of the country, both dated to the 5th century,[42] from England in the 5th and 6th centuries,[43] and Egypt c. 250 BC.[44] The earliest skeletal case is found in India, dated to 2000 BC.[45]

Fig 66 Man affected by leprosy from the early 20th century, his face is nearly beyond recognition.

Fig 67 Changes in the feet of a young individual buried at Sigtuna. (Photo Staffan Hyll, Riksantikvarieämbetet UV Syd)

Apart from airborne infection between people, certain researchers believe that the bacteria could also spread via damp earth. For example, it is likely that lepers, poor people living close to river banks and in harbours, walked barefoot on damp soil.[46] Leprosy is thought to have come to Europe with ships and along trading routes from India and Egypt. The disease is caused by the bacteria *Mycobacterium leprae* and is also called Hansen's disease after the man who discovered the bacteria.[47] It exists in different forms, but the two extreme forms are called lepromatous and

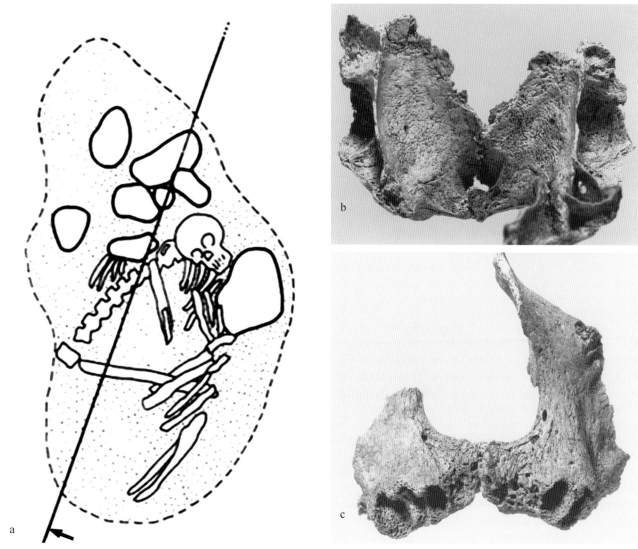

Fig 68a–c Oldest case of leprosy in Sweden, found at Sannagård in Halland. a. The position of the women buried in the mound. b. Superior view of the nasal surface showing the porosity. c. Frontal view note the resorptive changes of the nasal spine and the alveolus to the front teeth of the upper jaw. (Photo Lars Westrup, Kulturen)

tuberculoid leprosy. The former is progressive while the latter is less aggressive.[48] Both forms leave traces in the skeleton, traces which in many cases can be identified by osteologists. Leprosy is not very infectious; one has to be near lepers for a long time or on repeated occasions, and it is more common for children than adults to be infected. Having caught the infection, it can take 5–20 years before the disease develops and causes symptoms.[49]

A closer study of the occurrence of leprosy in Scandinavia in the Viking Age and Early Middle Ages, the time before the coming of leper hospitals, shows that the disease quickly gained a foothold in several places. In the earliest days of Lund in Skåne we have the largest reported accumulation of leprosy cases buried in ordinary church graveyards. Most of them have been noted in the oldest cemetery, Trinitatis, dated 990–1050 (Fig. 69), but cases also occurred at Kattesund (K3), 1050–1100, St Andrew's, St Michael's, and St Martin's cemeteries in their oldest phases, 1050–1100.[50] In the rest of Skåne a large number of cases have been found as well, for example at the oldest graveyard in Löddeköpinge, in Helsingborg, at the graveyard in Tygelsjö, and at the Viking Age burial place in Fjälkinge. In the rest of Sweden we can note early cases at the burial grounds at Kopparsvik on Gotland, Skämsta in Uppland, and among the early burials at Klosterstad in Östergötland. Somewhat later cases are found in Örberga and Linköping, both in Östergötland. Besides at Skämsta in Uppland, leprosy has been noted in Sigtuna[51] and in the graveyard at Västerhus in Jämtland (Fig. 70). The people with leprosy were buried in ordinary graveyards because there were no leper hospitals yet. The

Fig 69 Plan of the Trinitatis churchyard, red dots represent cases of leprosy.

oldest leprosorium in Scandinavia was established in the mid-12th century in Lund.[52]

The early cases of leprosy in Lund have been found on the edges of the graveyards. This may indicate that lepers belonged to the group of people who could not choose a plot for their graves. As regards the few cases at Fjälkinge and Kopparsvik, their placing does not say anything about their social status. At the graveyard in Löddeköpinge they are all buried in the churchyard that belongs to the earliest church (Fig. 71) but, unlike in Lund, they were found both near the church and in the periphery of the churchyard (Figs 69 and 71).

A number of skeletons with signs of leprosy have undergone strontium analysis. The results show that none of the lepers came from the place where they were buried. The strontium values of the lepers vary greatly between the different places, however, ranging from 0.708 to 0.738, which indicates that they did not have the same geographical origin (Appendix Table 2). There is a great deal to suggest, however, that both non-local people and those afflicted

Fig 70 Cases of leprosy from different burial grounds in Sweden. The red dots represent the oldest case from Halland, black dots represent the Viking Age cases, green dots the period cases dated to AD 1050–1100, and purple dots the early Middle Ages.

with a disease that was new to the community, as leprosy was, had the same access to a burial place as everyone else. Also, all the evidence shows that their graves had the same form as other people's and there were no differences in grave goods.

Health in Viking Age society

The overall picture of health in the Viking Age shows that the period did not differ much from, say, the later Middle Ages. Certain infectious diseases such as syphilis had not yet reached Scandinavia or Europe. On the other hand, as this study demonstrates, leprosy began to gain a foothold and spread to different parts of Scandinavia. Where it came from is hard to ascertain, and strontium analyses indicate that there may have been several different places of origin. Judging by stature, the general state of health appears to have been good for both men and women. A comparison with later times shows that the average stature was the same in the Middle Ages and at the start of the 20th century. Average life expectancy is harder to judge. It is known that osteological assessments of age based on archaeological skeletal material do not always give a correct picture of how old people became. The population studied at Fjälkinge shows that quite a few people became old even in the Viking Age. Certain skeletons found there were so osteoporotic that they were as light as paper, and simultaneously showed typical old-age fractures. The fractures were healed, which indicates that old people were taken care of.

Other clear signs of care for people with disabilities are the three dwarfs, from Skämsta and Kopparsvik. Not only did they have the handicap of being much shorter than everyone else, they also suffered extensive joint problems. Although the children were different at birth, the fact that the dwarfs reached middle age shows that they were cared for and had the opportunity to become as old as other people. The same applies to the women at Vannhög who had the Klippel-Feil syndrome.

Healed fractures, albeit sometimes healing in the wrong position, also reveal that people were cared for. During the healing process the broken bone meant that the person could not be active and needed help. This was also the case for those with joints that were "worn-out" or deformed by rheumatic pains. Several of them were additionally very old.

Although the picture of Viking Age society is closely associated with violence and conflict. Evidence of fractures and injuries incurred in combat is not more common on the skeletons than in later times. It is particularly rare to find injuries than can be linked to fighting with sharp weapons such as axes, *scramasaxes* and swords. There may be an unknown figure here, however, since this study is based on the results of inhumation graves. It is impossible to say anything about violence in society based on cremation graves since the bones are too fragmented.

At the start of the chapter we asked whether we can detect signs of care of the disabled and the sick, and whether there was acceptance for people who were different. Based on these results, there is much to suggest that the sick and the disabled were looked after. That many people not only reached adulthood but also old age is another sign that many enjoyed good health. As for the Viking Age as a society where violence was common, this study has not been able to confirm that picture. There is nothing to suggest that the frequency of fractures or weapon related violence was greater than in, say, the Middle Ages; if anything, it was lower.

4. Health and care for the frail

Fig 71 Plan of the Löddeköpinge churchyard, red dots represent cases of leprosy

5

Markers of identity?

As the social being that humans are, we need to show belonging; no one wants to be alone or left out. Those who do not feel that they belong in the crowd, in various ways and for different reasons, still want someone to have a sense of community with. Everyone at some time in life has noticed a sign of recognition in another person. Our social identity is created in childhood when we look at people and identify with some and distance ourselves from others. Interaction between humans often takes place with the aid of different symbols. To signal our belonging as regards sex, age, group, ethnicity, ideology, or lifestyle, we use things like clothes and body decorations in different forms. Some of these decorations are chosen to stay forever, others are used only on special occasions.

Body decorations of different kinds have occurred everywhere on earth, in all cultures and far back in time. They can be temporary adornment for everyday use but also for special situations. Beads made of animal teeth or shells are among the oldest ornaments we know of.[1] Presumably people also decorated their bodies with paint and used their hair in various ways to create hairstyles to highlight their appearance.[2] Some body decorations are permanent, and these include tattoos. The oldest known tattoo has been found on a mummified man found on a mountain top on the border between Austria and Italy. The mummy is called Ötzi (also known as 'the Ice Man') and he is 5300 years old. Studies show that Ötzi had tattoos at several places on his body,[3] and these were located at points where it has been found that his joints had caused problems. In this case the tattoos were probably done for medical purposes. On Ötzi's body there were tattooed lines on the wrists, on the left and right side of the lumbar vertebrae, a cross-shaped mark on the inside of the right knee, and three groups of lines on the left calf, a cross on the outside of the left foot, on the front of the right ankle joint, and beside the inner and the outer ankle. X-rays have shown that at several of the tattoo locations he had age-related lesions in the skeleton. As for the group of lines on his calf, they could have been intended to relieve muscular pains and cramp. We do not know how often he walked in the mountains.[4]

Other types of body decoration are incisions in the skin which result in scar tissue in different patterns, and piercing in all its forms.[5] Several body decorations can also occur simultaneously on the same individual, such as piercing and tattooing. Artificial deformation of the skull also occurs, for example, in South America and in Europe as well. In China women's feet were deformed through binding to give "lotus feet". Deformation of skulls and feet was closely linked to group belonging, and the process started early in childhood.[6] Increasing the number of neck rings in layer on layer in order to make the neck seem longer was likewise started early in childhood.[7] The number of cervical vertebrae is the same but the weight of the rings presses the collarbones and the ribs down. Certain permanent body decorations, such as skull deformation, can be detected in archaeological skeletal material.[8] For several hundred years, from the 16th to the 19th century, it was the fashion for both women and men in Europe to wear a corset. For men the purpose was to make them look impressive and for women to show off a narrow waist.[9] This fashion was anything but good for health, since it caused breathing difficulties and other problems.[10]

Studying the development of body modifications through history is difficult, since many types of lesions are in soft tissues such as skin and hair and, thus, can only be examined on mummies. Since there are very few preserved cases of body decoration from prehistoric and historical times, our idea of their age must be extremely limited. A form of body modification that is as widespread geographically as tattooing and piercing involves changing the look of the teeth. Dental enamel is the hardest substance in the body

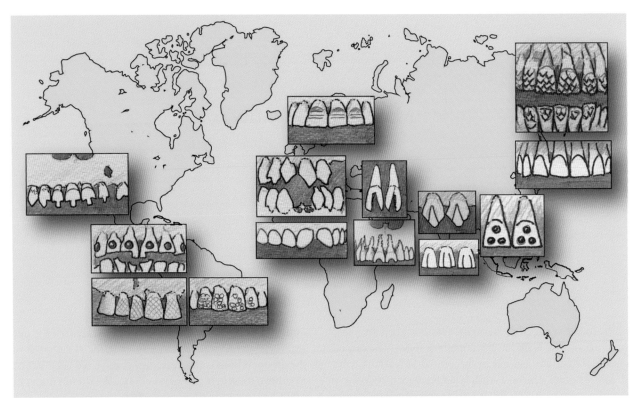

Fig 72 Variation of modified teeth from different parts of the world.

and is preserved even after the skeleton has decomposed and disappeared. Since the end of the 19th century scholars have noticed and documented different forms of dental modification all over the world (Fig. 72).[11]

Studying dental modification is thus a way to obtain knowledge about human behaviour and how it can be linked to different cultures, and the possible underlying causes. Certain ideas behind the phenomenon of filed teeth seem to be recurrent, irrespective of where in the world it is found, and scholars have often viewed it as an aesthetic and cultural expressions.[12] Some researchers who have studied the phenomenon of modified teeth in Central and South America have suggested that it is associated with social status, that is, the elite in a society.[13] Others argue that there is little or no evidence for a link to social status.[14] In studies of the phenomenon in the Philippines, however, it has been observed that filing of teeth in different forms is far more common than decorations in the form of inlays of gold and jewels.[15] Moreover, it has been noted in graves where the teeth of the individual were decorated with gold or jewels that there were also more prestigious objects present, that is, yet another indication of high social status.[16] Skeletal material from Central America displays great variation as regards the different types of modified teeth, even within adjacent areas. A study of material from Belize suggests that it is an expression of social and political affiliation.[17]

The oldest known case of dental modification in the world comes from India and was found at burial site no. 4 in the Bhimbetka rock shelter. Based on the stratigraphy of the cave, it was judged that the individuals, five in number, lived about 8000 years ago.[18] From Africa there are examples dated to 4500–4200 BC.[19] In certain countries in Asia and elsewhere, permanent dental modification is still performed today. In Europe too, it has been a fashion in the last decade to acquire a dental ornament in the form of small glittering stones or figures in gold.

In most cultures throughout the world dental modification is seen in both men and women over the age of 16. In certain places it is more common in women, in other places in men.

Filed grooves on the teeth

What reason can there be to discuss body modifications during the Viking Age? The answer is that the first osteological observations were made at the end of the 1980s, showing that people in Europe also changed the appearance of their teeth. The modification is described as follows: "strange marks on the teeth, as if someone had carved or cut horizontal grooves in them with a knife". A preliminary survey showed that these were horizontally filed grooves, especially on the upper front teeth, although canines and occasionally premolars were also involved.[20] Continued studies have shown that the marks also occur on teeth in the lower jaw.[21] The first three cases observed all had a crescent-shaped groove uppermost on the tooth, followed by one or two narrow nicks (Fig. 73). Continued study has

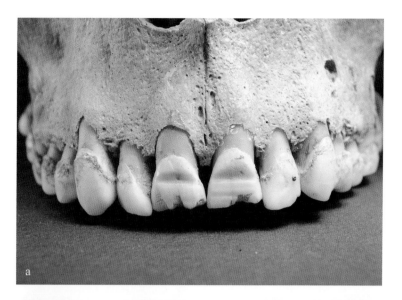

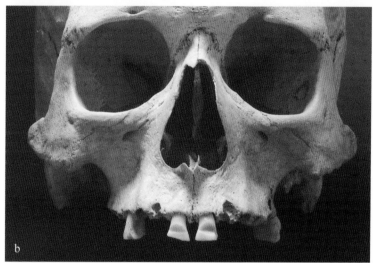

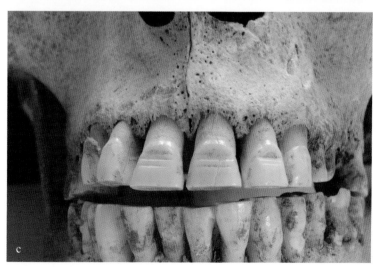

Fig 73a–c The first three cases of filed teeth that were found in Sweden; Trinitatis in Lund, Fjälkinge and Vannhög. All had a crescent-shaped groove on the tooth, followed by one or two narrow nicks. (Photo Staffan Hyll)

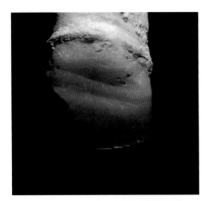

Fig 74 Not all filed furrows are horizontal, there are cases with diagonal furrows. (Photo Staffan Hyll)

Fig 75 In some cases, the filed furrow only extends over half the tooth. (Photo Staffan Hyll)

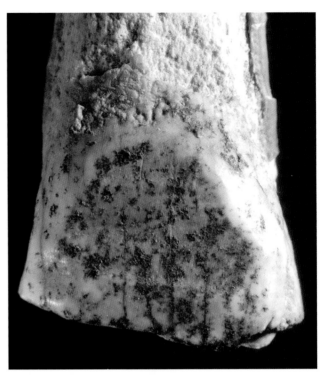

Fig 76 In several cases the technique used is to file the slightly convex tooth to achieve a flat surface, into which the horizontal grooves would be carved. (Photo Staffan Hyll)

Fig 77 Map of Sweden showing the distribution of male skeletons with filed teeth

demonstrated that the form of the grooves varies. Certain teeth do not have a clearly shaped crescent but only several narrower grooves. In some cases the grooves are very thin, in other cases so deep that they are a hair's breadth from reaching the pulp at the centre of the tooth.

Although the grooves are mostly horizontal, occasional examples are diagonal (Fig. 74), and in some cases they extend over just half the tooth (Fig. 75). The reason for the latter is the position of the teeth in the jaw, that is, a tooth that is only half-filed is partly covered by another tooth. Many have the modification high up on the tooth, starting where the tooth is convex. In several cases we note that the technique applied to the filed teeth involved first filing the usually slightly convex tooth to give a flat surface in which the horizontal grooves (Fig. 76). It was thus when the tooth was filed flat that the crescent-shaped surface arose, and those who had convex teeth acquired a more distinctly crescent-shaped surface. In certain individuals the dental modification stopped at filing the surface of the tooth flat. Some individuals have modifications which are so like each other that if teeth were swapped between two individuals it would be difficult to detect the difference. The very first survey showed that it was only men who had filed horizontal grooves in their teeth,[22] and subsequent studies have confirmed that this was a male practice.[23]

What is the geographical distribution of the phenomenon? The first survey showed that filed teeth were noted in 23 men, at five different places in three provinces in Sweden.[24] Today's picture is different. We can now state that male individuals with filed teeth have been found in five provinces, at 25 places, and in 132 individuals within Sweden's present-day borders (Fig. 77). The number is now more than five times greater. The two new find places are in Västergötland and Uppland. In Västergötland a man with filed teeth has been found at Varnhem, in the Viking Age part of the burial ground.[25] From Uppland there are several cases from different places such as Birka on Björkö,[26] and at the graveyards in Sigtuna,[27] in Bollstanäs and in Gnista. The latter are places north of Stockholm.[28] Cases have also been found at Tuna in Alsike, south of Uppsala.[29] Already in the first survey, it was evident that Gotland dominated regarding the presence of individuals with filed teeth, and the number of newly discovered cases on Gotland has now grown even more.[30] On Gotland it is above all at the burial grounds at Kopparsvik in western Gotland, and Slite on the east side of the island that the phenomenon is dominant, but it also occurs at other burial grounds. At certain places there is only a single skeleton, while at others there are so many that it can be regarded as a common phenomenon. At Kopparsvik and Slite the frequency is 37% and 64% respectively. The results also show clearly that the custom is more prevalent in eastern Sweden (Fig. 77).

Young, old, short, and tall

In order to come closer to an answer about the underlying causes, we have conducted a closer study of the age of the men with filed teeth. If the modification was done in adolescence it is conceivable that, like tattooing, it was done because of group pressure. If it occurs in teenagers it could be an initiation rite: a boy's entry into adult life. The results show that filed teeth are found in adult individuals of all age groups and proportionally in relation to the age distribution of the group as a whole. The filing can of course have been done on young adults, as it lasts for life. Yet the phenomenon has not been observed among any men under the age of 20. In other words, there is nothing to suggest that it is a rite of passage. Scanning electron microscopy of some of them shows that they had eaten after the filing was done.[31] There are also some with calculus in the filed groove, indicating that the filed furrows were produced well before death.

As has been mentioned, and as several of the pictures show, there is variation in the number of grooves as well as the depth of the grooves. No association can be detected between the number of grooves and the individual's age. The question is, were the grooves made one after one on different occasions? In cases where grooves occur over several teeth, they run neatly in parallel, which indicates that they were made at the same time. In other words, there is nothing to suggest that the number of grooves increased over time.

Can stature say anything about the men with filed teeth? If they were taller than average it could be envisaged that they belonged to a group within the elite. We have already mentioned that there is a link between stature and living conditions. If these men are instead shorter than average the interpretation could be that they belonged to a group whose life was not so good. Some scholars have suggested that the filed marks could be a sign of slavery, which in that case might indicate shorter stature than average since they did not receive the same amount of food.[32] The results, however, show a stature distribution corresponding to that for the general population (Fig. 78). Among the men there were both some who were short and some who were extremely tall.

Buried like other people?

How were the men with filed teeth buried? Is there anything different about the mortuary practice? It turns out that their bodies were buried in a wide range of ways. Some were given just a simple hole in the ground, while one man (grave 4) in Trelleborg was buried under a mound. For individual 496 in Birka a wooden chamber was built. The men with filed teeth are buried in as great a variety of ways as other people. As for grave goods and other personal objects in the graves, there is likewise great variation. Some have no grave goods at all, not even costume accessories. Others were dressed in fine clothes, for example, the man at Gällung in Väskinde parish on Gotland.[33] The grave at Gällung (grave 8) is an equestrian grave in which a man, his horse, and equipment were found. The man's body was buried in a pit that was covered with a rectangular stone packing. His equipment, which consisted of a short sword, a *scramasax* in a belt, a suspension ring for a scabbard, a sturdy comb, and a large ring brooch. On the right side of the man, from the lower leg down, a horse was placed. Already during the excavation it was noted that there were black marks on the man's skull and upper arm. The marks on the back of the skull turned out to be traces of a crossed band of silver, and on the upper arm they consisted of a silver decoration (Fig. 79).

The interpretation is that the man wore a cap with silver bands and a short-sleeved jacket of thin material to emphasize the silver decoration. Another well-furnished grave is number 496 north of Borg in Birka, which has already been mentioned above. The man had been given more personal objects such as tablet-woven bands of silver, pendants with mica discs, a belt, a buckle, and a comb. He had also been given a fragment of an Arabian silver coin, a Samanid dirham minted between 900 and 1000. Also in the grave were a whetstone, a knife, an arrowhead, a shield boss, a spearhead, a sword, riding equipment such as a stirrup and a frostnail, part of a pot, and three weights.[34] The metal embroidery, according to Hägg, displays oriental influence.[35] At both Birka and Gällung the men were buried with weapons. One may then ask how common it was for those who had filed teeth to be buried with weapons.

Weapons were found in the graves of only seven of 132 men with filed teeth, and these were distributed among the following burial grounds: three at Kopparsvik (graves 118, 235, and 280), one at Gällung (grave 8), one at Slite (grave 25), and two at Birka (graves 496 and 886). There is thus nothing to suggest that the filing of the teeth can be

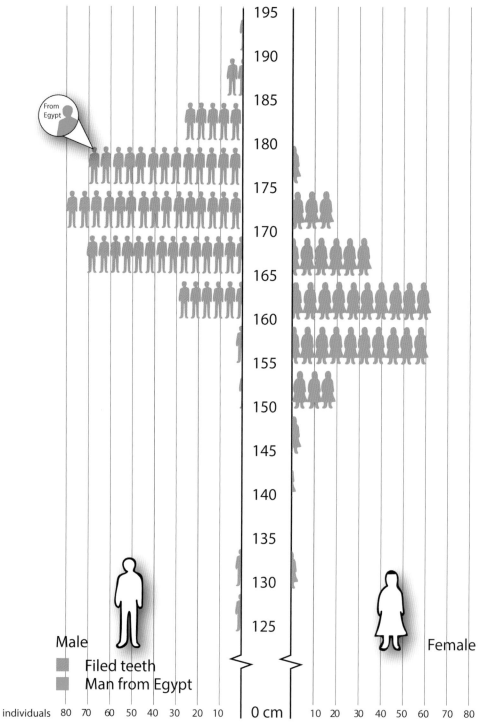

Fig 78 Variation in stature among the individuals with filed teeth. The orange figures represent the men with filed teeth and the green figure a man from Egypt.

directly associated with the role of warrior. However, not everyone who belonged to some form of group with the duty of defending and attacking was of necessity buried with weapons. Since a few of them were actually buried with weapons, it is interesting to see how many men with

filed teeth show injuries from weapons or blunt instruments which could be a result of combat. In the total material we have hitherto noted five men with filed teeth and weapon injuries. In three of the cases the men incurred mortal blows, at Vickleby on Öland (grave 1028), at Kopparsvik (grave

Fig 79 Black marks on the back of the skull, representing traces of a crossed band of silver. In the top right corner of the image we see a bridle, underneath it threads of textile and small silver rings and finally at the bottom filed teeth.

154), and in Birka (grave 138). In addition there is a man with fractures at different places on his body which could have been caused by a fight with blunt weapons (grave 4 Vannhög). One man was executed by beheading, at Bollstanäs (grave 29/f59). None of them was buried with weapons. If we compare the occurrence of weapon injuries on those with filed teeth with other individuals, we find no difference.

As for the position of the body in the grave, we can observe that most of the men were buried on their backs, but there were also some who were placed face down. This applies especially to the individuals from Kopparsvik,[36] but there is also another example from a grave at Bollstanäs in Uppland.[37] Nor is there anything to suggest that the location in the cemetery differs from that of other people, or that those with filed teeth are found in any distinct grouping in the burial grounds.

Did the phenomenon of filed teeth occur throughout the Viking Age? The majority of the graves at the studied sites have been dated by means of finds, but in some cases dating has been done by radiocarbon, and the results indicate that the oldest graves are from the start of the 8th century and the most recent are from the middle of the 11th century, which appears to agree with the datings of the burial grounds as a whole. There are however variations in dating between the different sites. One of the early cases is from Bollstanäs AD 835±110.[38] One of the graves at Vannhög (grave 4) in Trelleborg likewise shows the early occurrence of the phenomenon, with a radiocarbon date of AD 783±71. Among the later cases that we know of, the man with the money bag at Slite (grave 8) is dated to AD 1120±59. The men with filed teeth are buried in different places at roughly the same time, so we may ask what this means. Did the idea arise at different places at the same time independently, or was there a shared place from which they all originate?

Was Gotland the gathering point?

The men with filed teeth were buried at different geographical locations, and the natural question then is whether they were originally from these places. Strontium analyses have been performed on several of the men in an attempt to shed light on the matter. The vast majority of the men with filed teeth have hitherto been found in graves on Gotland. One hypothesis is that the custom had its origin on the island, and that the men with filed teeth whom we find in other places than Gotland were Gotlanders who had moved from the island. If the custom is Gotlandic, according to another hypothesis, those who were not originally from Gotland could have been there to get their teeth filed; after visiting Gotland they either went home and died there or travelled on and died somewhere else, and were thus not buried on Gotland. A third possibility is that the modification of the teeth was performed in other places and that Gotland, for some reason, was a gathering point for something that the filed marks represent. The dates have now shown that they may have been widespread as early as the end of the 8th century, that is, at the start of the Viking Age.

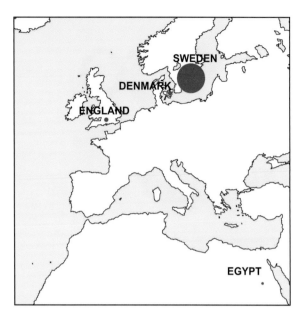

Fig 80 Map of Europe with distribution of men with filed teeth. Note the high frequency in Sweden compared to Denmark and England.

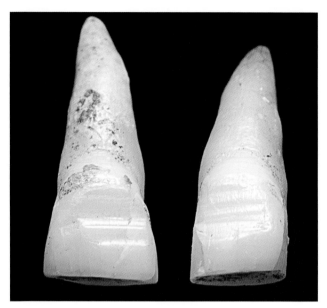

Fig 81 Filed teeth from Dorset, England, with a crescent shape and another stripe on the front teeth.

As we see from Table 3, the strontium analyses indicate that of 20 men with filed teeth who were analyzed, 13 may have had their origin on Gotland (Appendix Table 2). The rest came from other places. At the burial grounds in Skåne it can be stated that the man who was buried in Lund (Trinitatis) has a value indicating that he was not originally from Lund, but could be from south-west Skåne or Gotland. At Vannhög there were two individuals, the older one of whom (the man buried in the mound) was not local, but he may have been from Gotland. The younger man may have been of local origin or come from Gotland. The man in Fjälkinge was not from the place where he was buried. He could have been a Gotlander or from south-west Skåne. For two of the individuals on Öland, one buried at Vickleby, strontium value 0,708, the other at Hulterstad, the strontium results show that the former could be from Öland while the latter cannot have been from present-day Sweden (Table 3).[39] Two of the skeletons from Birka have undergone strontium analysis and neither of them is from Birka. One of them, however, may be from the Mälaren region while the other shows values corresponding to those from Öland, for example. The conclusion is thus that some of the individuals with filed teeth who have been found at places outside Gotland may have been Gotlanders abroad, while others were originally from other places. The custom might be Gotlandic in that sense that many of those that got their teeth filed had it done on the island.

We know from international literature that there were special practitioners of dental modification.[40] In view of the fact that the phenomenon occurred in Scandinavia over a couple of centuries, there must naturally have been multiple people performing it. The large number of men with filed teeth on Gotland suggests that the island exerted a particular attraction that can be linked to this phenomenon. Apart from the cases of filed teeth in Sweden presented here, it should be mentioned that there are a further three cases in Europe. Two of them were found at Galdegil on Fyn in Denmark,[41] and one case in Dorset in England (Fig. 80).[42] The marks on their teeth and the dating agree with the evidence from Sweden. It is not strange that there were two in Denmark; if anything, it is surprising that more have not been found. In Sweden, admittedly, we can clearly see that the phenomenon is primarily associated with the eastern side of the country. As for the man in Dorset, strontium analyses of a selection of the 50 men who were executed there show that they were not local but probably from Scandinavia.[43] The man from Dorset has a crescent shape and yet another stripe on his front teeth (Fig. 81; skull number 3736).[44]

A Nordic custom or inspiration from elsewhere?

As already mentioned, the phenomenon of modifying one's teeth is a custom that goes back a long way, and it is scattered all over the world. The different types of modification vary depending on where in the world they occur. Certain forms are unique while others are very similar to each other and in some cases almost identical (compare Fig. 75 from Gotland and Fig. 81 from Dorset). They both had filed furrows on just half the tooth.

Cultural phenomena, of course, can arise at different places without any connection between them. The type of modification displayed by the men in Scandinavia, with horizontal file marks on the teeth, has not yet been noted anywhere else in Europe except in Sweden, Denmark, and England (Fig. 81). On the other hand, horizontal grooves have been observed on the teeth of individuals in

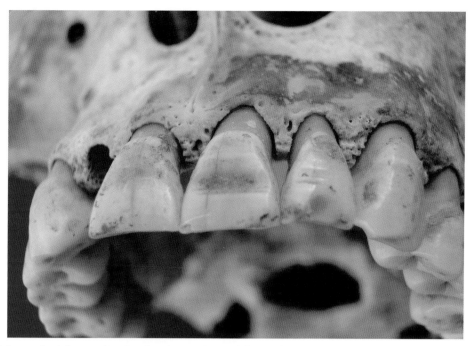

Fig 82 Filed teeth from a burial ground in Oxyrhynchus in Egypt. (Photo Bibiana Agustí. Archaeological Mission of Oxirrinc, director Dr. Josep Padró, University of Barcelona)

graves in Illinois, Arizona, and Georgia in the USA.[45] The individuals with these modification belonged to the groups who used the mounds at Cahokia, Illinois, but there were also individuals from the Lamar culture in Georgia. Among them there were horizontally filed teeth in both men and women. What distinguishes the Scandinavian cases from the American ones, however, is that the former were filed to give a flat surface before the horizontal grooves were filed. The approach is not the same.

Marks identical to the Scandinavian ones have, however, been found in a completely different place, namely Egypt, at a burial ground in Oxyrhynchus in the province of Al Minya (Fig. 82).[46] This is a man buried in the upper strata of Tomb 16 of the Saite period. The filed grooves on the man's teeth have exactly the same appearance as those in the Scandinavian material. The man was a relatively young individual, 25–40 years of age. The archaeological context indicates a dating of the grave to AD 300–600. Written sources show that the Muslims conquered Byzantine Egypt in AD 641. Christians nevertheless continued to be a significant share of the population up to the 11th century. The man with the modified teeth was buried together with 39 other individuals, the majority of whom were children.[47] His skeleton was found among the last to be buried. We have radiocarbon-dated the skeleton to AD 717±38. Although the burial custom for the man indicates that he was local, we thought it appropriate to perform a strontium analysis to ascertain the geographical origin. The result showed a value of 0.707, which agrees with the baseline derived from animals from the locality.[48] Compare the filed furrows on the teeth of the man from Vannhög with the man from, Dorset and the man from Oxyrhynchus, they are quite similar (Fig. 83)

The question then is how this occurrence of the filed teeth in Egypt should be interpreted in relation to the Scandinavian examples. In Africa modified teeth in skeletal material have been studied since the start of the 20th century.[49] The custom was noted by the famous explorer Livingstone in the 19th century.[50] Studies performed hitherto have shown that different forms of tooth modification are found in different places in Africa with everything from chipping to filing, whereby parts of the corners of teeth are reshaped.[51] As far as we know, the type of dental modification observed in Egypt has not previously been noted in Africa.

Is there any link between dental modifications in Viking Age Scandinavia and the man in Egypt, or did they come about without any knowledge of each other? In Egypt we know at present of only one individual and in Scandinavia more than 130. Viking Age literature has several sources indicating that people, perhaps chiefly men, travelled extensively to places both near and distant. To simplify, those living in present-day Denmark travelled west and south towards England, France, Spain, and into the Mediterranean from the west. From present-day Norway the Vikings voyaged north towards Iceland, Greenland, and even North America. Viking Age people living in what today is present-day Sweden are best known for their eastward journeys towards present-day southern Russia, down to the Black Sea and the Caspian Sea. They also travelled to the Byzantine Empire and present-day Greece and Turkey, and as far east as the areas now known as Afghanistan

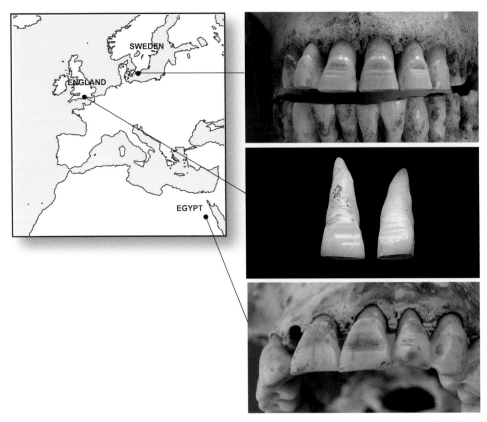

Fig 83 Comparison between the filed furrows on the teeth of the man from Vannhög, Dorset and Oxyrhynchus, they look quite similar.

and Uzbekistan.[52] As far as is known today, however, the journeys to Byzantium mostly took place in the 10th century, that is, after the men in Scandinavia had begun the custom of modifying their teeth. Should the case in Egypt be viewed solely as a coincidence, a custom arising independently in two places, or is it possible that there was some contact, some way the idea could spread?

An interesting observation in this connection concerns grave 496 in Birka. For men in Scandinavia to be inspired, however, it was not necessary for them to have been in Egypt or for Egyptians to have been in Scandinavia. The custom may have been passed on through other channels as a result of repeated encounters at mutual trading sites. In other words, the custom could have moved in the same way as finds in different forms. It thus cannot be ruled out that Norsemen were inspired by other people. That there is only one individual in Egypt may be because it has only just been discovered, and when this publication appears it may make other scholars aware enough to look out for more cases. Naturally, one may wonder whether it might be a Scandinavian custom that spread to Egypt, but hitherto the radiocarbon dates indicate that the Egyptian example is slightly older.

In recent years archaeological discoveries have shown that Scandinavians travelled east at an early stage. We saw above that in Salme on the island Saaremaa in Estonia, two boats from the 8th century have been found, containing a total of forty men who had met a violent death.[53] None of the men buried in the boats, however, had any file marks on their teeth (Raili personal communication). Nor have any cases been reported from the graves in, say, Hedeby and Dorestad, or the graves at Bodiza in Poland.[54] Although almost thirty years have passed since the first case of filed teeth was discovered, and despite fervent contacts with colleagues all over Europe, the cases outside the present-day borders of Sweden are vanishingly few. Are filed teeth absent elsewhere in Europe simply because there are no cases, or have they just not been noticed? We believe it is the latter.

Why file grooves in teeth?

The last question we must ask is why filed grooves in their teeth. We should once again turn to the Arabian chronicler Ahmad ibn Fadlan's account of his meeting with the Rus, that is, the Vikings. He describes their appearance as follows: they had ruddy skin, blond hair, and tattoos from the fingernails to the neck in dark-blue or green patterns resembling trees or other figures.[55] He is not the only one to note the body adornment; Adam of Bremen (died before 1095), a churchman and historian, writes the following about the people living on an island named Samland, a peninsula in the Kaliningrad region: "These men are blue of colour, ruddy of face, and long-haired."[56] The question then is, were the filed grooves part of the body decoration, and if so what did the people wish to communicate? The filing of the teeth was a deliberate act, as indicated by its prevalence, and also by the fact that men who have been found at distantly separated geographical places show such similar formations on the teeth that there are grounds to suspect that the decoration was done by a group of "craftsmen".

The question, of course, is whether they had their teeth filed solely for aesthetic reasons, or if it was an expression of belonging to a group. If we look back at the characteristics of the men with filed teeth, we get the following result. They have been found in different types of graves, in lavish chamber graves, in equestrian graves, but also in simple contexts where the grave was only a hole in the ground. In a very few cases we have noted that they died a violent death by someone else's sword or axe. The majority were buried on their backs, but we also find them among those who were buried face down. They are adults of all ages, but no children or teenagers. They do not differ in stature from other men at the time, nor did they suffer injuries and diseases more than others. They lived and died at different places in Scandinavia. As far as we know today, however, the majority were buried on Gotland, although the strontium values show that many of them were not born and bred on the island. What then was the attraction of Gotland? The Viking Age is closely associated with violence and raiding, but as we have seen, there are no clear signs that filed teeth are associated with a warrior group. At any rate, very few of the men with filed teeth have traces of weapon injuries. The proportion of the men who were buried with weapons is not large either. Nor are there any indications anywhere else in the world that modified teeth have anything to do with fighting groups.

Two of the men with filed teeth had fine bands of silver thread on their clothes. One of them was found in a chamber grave in Birka (grave 496) and is described as follows:

> The grave goods indicate a man of the highest social standing with passementeries [trimmings] that, according to Hägg,[57] have been identified as symbols of rank related to Byzantine court dress. The weaponry consisted of a spear, a shield, one arrowhead and a sword with a bronze chape decorating the sheath. The presence of horse-gear and a platform for the remains of a horse enhance the high status of the interred. The grave also contained weights and Islamic silver coins. The burial is dated to the first half of the tenth century. The overall picture is that this is the burial of a ruler or an influential individual at the highest level of society and that he has been buried with symbols of his position, wealth and office.[58]

Fig 84 The hoard from Hägvald, Gerum on Gotland. It weighed 3.5 kg and contained 1298 silver coins, four spirals and 31 pieces of silver found in the vessel. (Photo Fornsalen, Gotland)

The other grave is an equestrian grave at Gällung, Väskinde parish on Gotland. His equipment consisted of a short sword, a *scramasax* attached to a suspension ring, a stout ring brooch, and perhaps a lance head. His horse was bridled and there was a sturdy comb for it. This man too had silver bands, on the skull, at the shoulder, and running halfway down the upper arm.[59] The question is, what type of group do these two men represent? Could they have been rich merchants who could afford to buy expensive clothes? People from Scandinavia, known as the Rus, played a major role in the development of trade networks in north-west Russia.[60] Can the many hoards found on Gotland be linked to the occurrence of filed teeth? Gotland is extremely rich in Viking Age hoards. There are around 750 silver hoards from the island but also hoards containing gold. More than 165,000 coins from Europe and the Arab countries has been found. Nowhere else in the World are so many hoards found. The treasure from Hägvald is a typical Gotlandic hoard from the Viking era, it is larger than average and almost intact (Fig. 84). It contained 1298 coins, mostly Arabic but also Volgaburgian, Byzantine, German Bohemian, and English. In addition, there were spirals and different pieces of silver and all was preserved in a large vessel that was placed on a flat stone and had another flat stone as a lid.[61]

6

Burial grounds designated for particular purposes?

Life goes hand in hand with death. Along with the bustling life in a village or town there is the silence of the graves. Burial places partly reflect conditions, circumstances, and events in the living community. In past societies, unlike today, it was most common for a burial place to contain the vast majority of those who took part in the local community up until their death, but there are also exceptions. From later times when written sources are available, we know that burial grounds were established outside the ordinary graveyards. One reason for the rise of new types of burial grounds was the scourge of epidemics,[1] and another was conflict and war.[2] Certain institutions also had burial grounds of their own, for instance the leprosy hospitals.[3] Criminals who had been condemned to death could not be buried in church graveyards; they were buried at the place where they had been hanged or beheaded.[4] These special burial grounds, depending on the underlying causes, can therefore have a different age distribution and sex composition than a farm cemetery, a village burial ground, or a parish graveyard.

The first question one should ask in this context is, how did a non-Christian burial ground arise? When was the first decision taken as to where a place of burial should be located? Did it happen when one or more farms were established in an area when the people who staked out the land and claimed it for other purposes also marked the last resting place for their kinsfolk? Were the social spatial structures and the boundaries of the burial ground determined when it was first used? There are many questions and it is difficult to obtain answers since the written sources from this time are scarce or almost non-existent. As this was a time when mortality was greatest among the smallest children, it is not unlikely that the first grave in many cases had to be dug when a newborn baby died. A newborn child, who had not been a part of the community, was perhaps the individual for whom the first grave marker was made. Burial grounds were not merely a storage place for dead relatives; they were also important in judicial terms.[5] A burial ground showed that the farm or the village had been in use for a long time. When the Viking Age was succeeded by the Middle Ages and the provincial laws were written (the section on land in the law of Västergötland and the law of Östergötland), we find that graves were of legal significance; in disputes over common land, for example, a village with burial mounds could prove that it had an ancient right to the land.

What age and sex composition can we expect to find in a burial ground that was used for a couple of hundred years by a more or less permanent population that was not exposed to major epidemics or wars that could have the greatest impact on certain age groups? Are there any examples that can serve as a model for what we can expect? Scholars who study the demography of hunter-gatherers in the past use anthropological studies of small groups living in the 20th century in conditions similar to those in the Stone Age. For the Viking Age population there are no such proxy groups. Instead we have to use written sources and death registers from rural parishes, although these are only from modern times – the 17th century to the 1840s.

Since the Viking Age skeletal samples discussed here are relatively small, we have tried to find historical materials with relatively small population sizes. We have found them in Norrland, Västergötland, on Gotland, and on Sjælland in Denmark. The two smallest parishes had 66–110 inhabitants, and the larger ones over 500. The study is based on data covering a 40–50 year period. The oldest register of deaths spans the period 1646–1688 and the most recent 1791–1840.

As the diagrams show, in three of the four parishes all age groups are represented (Fig. 85). The exception is Akebäck on Gotland, where no older teenager died. It is

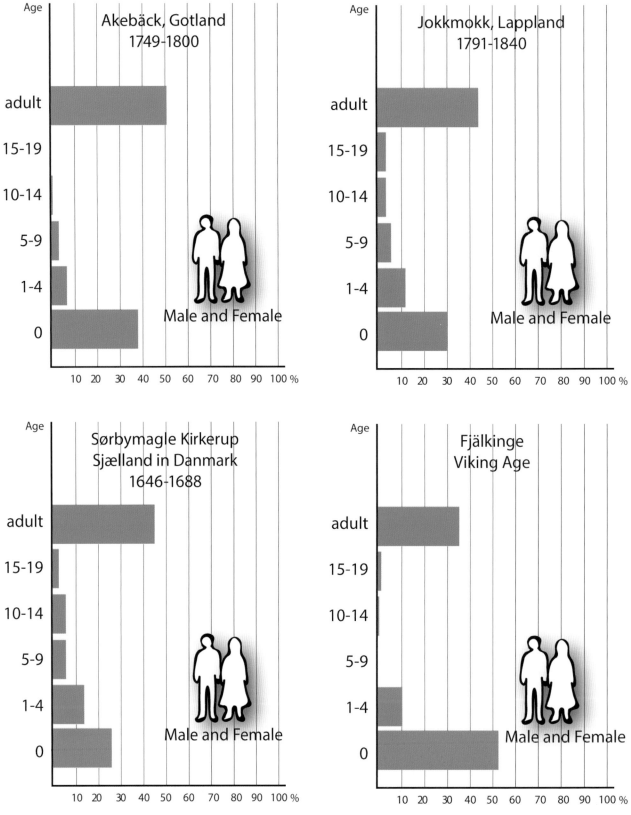

Fig 85 Age distribution of four historical parish registers from four different geographical areas. Except in Akebäck, all age groups are represented.

also evident that in the material based on written sources we note a higher mortality among the youngest children than in the older age groups. It is important to note that these are rural parishes not densely populated towns with a larger population where infectious diseases can be rife for a longer time. That means that, during normal circumstances, this demographic pattern is not that far from what we can expect from conditions during the Viking Age. In other words, a high frequency of children 0–1 years of age and also quite a number of old people.

The influence of Christianity or division into special areas?

Let us look at the demography of the burial grounds that show an unusual age composition. It is the rule rather than the exception that the proportion of infants in prehistoric burial grounds is very low. On the other hand, older children of all ages are often found. Another age group that is mostly also significantly under-represented in comparison with historical times is the elderly, that is, individuals aged over 60. Several of the burial grounds in this study show this demographic pattern. One site, however, differs noticeably in this respect, the cemetery at Fjälkinge. There the proportions were reversed. Infants accounted for just over half of the burials (Fig. 20), while the elderly made up a quarter of all those who died in adult age. Women in particular lived to a considerable age.

If we ignore the proportion of infants and look more closely at the distribution of the other age groups, we find other discrepancies in the Fjälkinge material, namely, that there are no children aged 4–11 (Fig. 20). The sex distribution among the adults is equal. Preservation conditions for bone on the site are very good, as is also evident from the fact that bones of very small infants and foetuses are preserved. According to the archaeologist who excavated the site, the burial ground has not been totally excavated. The small-scale excavation by Strömberg in 1953 shows that four adult skeletons were found under the road beside the burial ground. One of them was buried face down. In addition, a group of seven graves was found on the other side of boggy land and a river to the west, and skeletons of six individuals were retrieved.[6] Five of them were children, including one infant, and the other four were larger children. The older children thus belong to the age category that is missing from the main burial ground.

For Fjälkinge we note that most individuals were buried separately, but there are also 14 examples of individuals being placed near each other in different ways, on top, alongside, or at angles to each other. In six of the cases these are small infants. Occasionally an adult and a child are buried together and, in these contexts, it was just as commonly a man and a child as a woman and a child. In one grave it was probably a mother and child. The child, which was not fully developed, was found between the woman's thighs. The woman was buried without a coffin; it is not a case of a foetus emerging from the body as the tissue decomposed. Although it is a somewhat unusual placing for the infant, it seems likely that the child was buried at the same time as the woman. The fact that such a large proportion of the bodies were laid together – and apart from other people – may suggest that the individuals in the groups were closely related. Since the individuals were buried on top of each other, but at different times, it is an indication that the location of the first burial was known. Perhaps there was some form of stone marking the grave.

What could be the explanation for the presence of so many infants at Fjälkinge? Earlier research claimed that not all were buried in prehistoric times, including the Iron Age, and that this applied in particular to infants, but that the coming of Christianity changed this.[7] The reason for the large number of infants in Fjälkinge would thus be that some of them, the latest to be buried, were Christians, and that a prohibition on the exposure of infants had come into force. The assumption that infants were not buried is based on statements in the Icelandic sagas about the exposure of infants, which are interpreted to mean that this was an accepted practice before the establishment of Christianity.[8] It was a high priority for the early church to eradicate this custom. Is it reasonable to believe that the exposure of infants was so common in the Viking Age? Some scholars argue that it was done for economic, social, religious, or biological reasons – the latter meaning that children with deformities were exposed. It is thought, however, that economic reasons predominated, especially in periods of famine.[9]

Scholars believe that Christian society was more virtuous and merciful because children, old people, and the sick were looked after. Archaeological excavations show that deceased infants and children were placed around the chancel in churches and along the walls of the nave, because of the Christian outlook on children as based on the words of Jesus in Matthew 18:3 and 19:14.[10] The question remains, however: is it really possible that people in the Viking Age killed and exposed infants to such an extent that it can be seen in the demography of the burial grounds? What is not mentioned is that, even in Christian society, there were restrictions which excluded certain groups; unbaptized children could not be buried in the graveyard and a handicapped child could not be baptized.[11] Nor could adults who were mentally ill and saw suicide as the last resort be buried in the graveyard. The same applied to people who had committed capital crimes.

To prove his hypothesis that an increased number of infant graves is an indication of the influence of Christianity, Mejsholm has compared two burial places in Skåne, one of them being Fjälkinge (AD 800–1050) and the other the graveyard at Kattesund church (1050–1100) in Lund.[12]

Both show a high proportion of infants and old people. Yet the frequency of infants at Fjälkinge is twice as high as at Kattesund. Not even the so often cited graveyard of Västerhus in Jämtland,[13] which is known for its high frequency of infant graves, has as many as Fjälkinge. If one wishes to claim that the reason for the high frequency of infants is that people had become Christians, it would have been more appropriate to compare Fjälkinge with, say, the Trinitatis graveyard in Lund (990–1050) which has only Christian burials. The demography at Trinitatis in Lund shows all age groups but the proportion of infants is very small, unlike at Fjälkinge. A salient point in this connection is that the Trinitatis graveyard, unlike Kattesund, has not been totally excavated. The area around the chancel, where small children ought to be buried, has however been excavated and no accumulations of infants were found there, not even traces of empty graves where the skeletons may have decomposed. That archaeologists note children's graves even if the skeletons are not preserved is shown by the excavation of another early graveyard, Drotten 6 just south of Trinitatis (1040–1100).[14] It is important to point out here that the children's graves at Fjälkinge to which most care was devoted are the pagan ones, most of which have coffins and are, moreover, dug deeper than those with a Christian orientation, with the body buried on its back.[15]

The other group that was frequently represented at Fjälkinge are old individuals, chiefly women. In osteological terms this is a difficult age group to identify as our methods are not sufficiently refined; people age differently and many individuals are, for that reason, assessed as being younger than they really are. If the skeletal material is badly preserved in addition, so that age-indicating criteria are missing, the risk of incorrect assessment is even greater. At Fjälkinge the preservation conditions were good, and many individuals provide an adequate basis for estimating the age; moreover, some of them had typical old-age fractures, for example, on the front pelvic joint and the lower part of the upper arm.

Does the occurrence or absence of certain age groups indicate that burial grounds like Fjälkinge came about in connection with specific circumstances or that people in certain times and among some groups practised some form of age division in burial grounds? Can the high proportion of infants be due to an epidemic among the very young? Not all the infants, however, were buried together; instead they lie in groups with others like small families. Moreover, the pattern of mortality is what can be expected even in today's society, namely, that most of the children who died did so very soon after birth.[16] In other words, there is nothing unusual. Another reason why there are so many infants may be that it was a society with an expanding population and a high proportion of newborn children. Since infant mortality at this time was high, one would expect to find many infants. Fjälkinge, however, is not the only cemetery from this time with a remarkable age distribution and an extremely large accumulation of small children.

The Triberga cemetery on Öland has been excavated on several occasions, and shows a composition that is at least as strange.[17] The cemetery was used for 400 years from the Vendel Period up to and including the Viking Age. Of the 24 children buried there, 18 (88%) were under the age of 1 year, and there were two children aged 1–2 years, one aged 7–8 years, besides which there were three adults. The majority of the infants are in the northern part of the excavated area. The burials at Triberga should be regarded as poor, with the exception of two graves. The accumulation of infant burials in northern part of the cemetery is interpreted by the archaeologist as showing that there was a spatial order in the cemetery and even that children were perceived as a separate group.[18]

Yet this is not a consistent pattern in all the cemeteries of Öland. At Alby or Folkeslunda, for instance, the children are scattered in clusters as at Fjälkinge. From earlier periods such as the Roman Iron Age it has been noted that there is a division into male and female zones and that the children lie in separate rows, for instance, the burial ground in Bjärby.[19] Can the cemetery at Fjälkinge be an example of separate burial areas for certain age groups, and can small groups of graves found in the vicinity of the cemetery[20] belong to the larger Fjälkinge cemetery? In one of the small excavated areas there were moreover several larger children.[21]

Market places and harbours?

A situation quite the opposite of Fjälkinge can be found in the two cemeteries at Slite and Kopparsvik on Gotland. At both sites there is an almost total absence of children and the sex distribution is highly uneven. At Kopparsvik the skewed sex distribution is moreover associated with spatial considerations, and the disproportion is greatest in the northern part (Fig. 29). Here men make up no less than 80% of the burials. In the southern half of the cemetery, by contrast, the sex distribution is more equal. The teenagers that have been found constitute a larger percentage in the male-dominated part. The presence of foetuses/infants can be explained as representing pregnant women who came to the place or who became pregnant when they were there. At the Slite cemetery we find few children, only three out of 30 investigated burials, and of these two are foetuses that lay with their mothers, once again pregnant women. No teenagers at all were found. The third child was assessed as being aged 7–14, probably in the first half of that interval. It must be emphasized that preservation conditions are good, and even though there are areas where the graves are closely spaced and have partly disturbed each other, there are large areas where the graves are widely separated and intact.

Yet another burial ground displays unusual distributions, namely, the one at Fröjel. Here we see a majority of females. Just as at Kopparsvik, the burial ground at Fröjel consists of two areas. In the southern part few graves have been excavated, but men clearly dominate (Fig. 35) and we find no children here. A larger share of the northern area has been excavated, where women dominate. The children that are represented are infants. Three of them lie in the same grave as women of fertile age, and there is every reason to assume that this is mother and child. At Fröjel there is also an early medieval graveyard east of the southern part of the Viking Age cemetery. Children of all ages except teenagers were found there, and infants made up about a quarter of all those aged under 20.

How should we interpret the skewed age and sex distribution at Kopparsvik, Fröjel, and Slite? Is this once again a sign of a divided burial ground? According to Westholm, the cemetery at Kopparsvik was used by the people who laid the foundation for what would later be the town of Visby. This would be farms in the neighbourhood, where both those who had the rights to the shore below the cliff and those who did not, had together formed a harbour association which established the burial place at Kopparsvik.[22] Yrwing, however, finds it odd that the harbour association did not bring their dead home for burial in the farm cemetery, and he believes instead that the people buried here belonged to special groups.[23] In the Viking Age the coast at Visby became important for the trading farmers of Gotland, for their collective journeys to the Mälaren area, Finland, the eastern Baltic, and Novgorod.[24] Thunmark-Nylén likewise thinks that the cemetery, especially the male-dominated northern part, belonged to a community performing special functions with men in the leading roles, for example, trading.[25] The large amounts of silver could have come to the island almost exclusively through trade.[26] Similar conclusions are presented in a recently published dissertation by Matthias Toplak,[27] comparing Kopparsvik with other burial grounds on Gotland.

The dominance of men at Kopparsvik has also led to speculation as to whether some military activity could have been carried on there. This thesis has not been developed or explained, however.[28] If it were a garrison, how is it identified in the grave material and who was it intended to protect? A garrison is a group of men equipped with weapons, tasked with maintaining order. With the exception of Viking Age burial places in Uppland, especially Birka, it is rare to find graves containing weapons.[29] Few of the men buried at Kopparsvik had weapons: only seven out of 242. Of these seven, six were buried in the southern part of the cemetery, where the sex distribution was even. The weapons here were *scramasax*, lance head, and axe. This is far fewer than in the graves at Birka.[30]

It was slightly more common to find weapons in the graves excavated at Slite, where they occurred in three of 22 male graves. Finds of Viking Age weapons, horse equipment, and equestrian graves do occur on Gotland. This is riding tackle of various kinds: spurs, stirrups, bridles and, in certain cases, a horse. Finds like these have been excavated in 44 of Gotland's 92 parishes. They are scattered over the whole island, both along the coast and further inland. Another sign of the existence of a *hird* would be traces of weapon injuries. At Kopparsvik, however, only two of the men display traces of violence involving sharp-edged weapons. Both died from their injuries and both were buried in the northern area, and neither of them was buried with weapons. On the other hand, there are several with healed fractures on different parts of the skeleton, for example, metacarpal, rib, ulna, and femur. Two of these individuals are women. These injuries need not have been incurred in battle, but could be the result of accidents or possibly fights. The overall picture at Kopparsvik thus gives no clear signs that the men belonged to a garrison or a group who suffered their injuries in combat, whether as attackers or defenders. At Slite, however, there are two men with minor weapon injuries on the skull, one of which had healed, the other was in the process of healing. Both were buried with weapons, one with a broadaxe, bearded axe, and spearhead, the other with a *scramasax*.[31]

If the male dominance in the burial grounds at Kopparsvik and Slite indicates a link to a harbour and trade, what can we say about the female-dominated cemeteries at Fröjel and Birka? For Birka it has been pointed out that the skewed sex distribution may be explained by the fact that it is easier to identify women's graves than men's based on the objects in the cremation graves.[32] At Fröjel, on the other hand, the skeletons are preserved and there too there are more women than men, with the reservation that the northern part has not been totally excavated. Can the explanation for Birka and Fröjel be that both places were not just trading sites but also permanent settlements, whereas the heavy surplus of men at Kopparsvik and Slite could perhaps be due to the fact that these were places that received a large number of boats in certain parts of the year and also had more men there to protect the routes to Gotland on both the east and the west sides? By all means we see some kind of division as regards both age and sex. The strontium analyses from the different places show that a considerable number of individuals were originally not from Gotland, but came there as teenagers or adults. The same also applies to Birka, and all three burial places in Skåne (Fjälkinge, Vannhög, and Trinitatis). Both inland and coast appear to have had relatively large-scale immigration. As regards Kopparsvik, it is also in the southern area that most non-locals were buried. Could it be that those who had work which required being close to the coast also had their burial places there, whereas those who worked with cropping and animal husbandry had their burial places at home on the farm? Matthias Toplak, who has studied both Kopparsvik and other burial grounds on Gotland has found a more equal sex distribution at cemeteries in the interior of the island, for example, the ones at Ire and Barshalder.[33] Although none of the burial places described here has been totally excavated, there is a distinct tendency for certain cemeteries in the Viking Age to have a division of individuals according to age and sex.

7

A time of many faces

What do we mean by a time of many faces? This study, together with others, shows that mobility was extensive among Viking Age society. Some people came from nearby areas, but there are also examples of long-distance migration. To what extent this migration was voluntary or coercive is hard to say. It was a time when many different faces met. Both those who travelled off to various geographically distant areas and those who stayed at home in Scandinavia had come into contact, via merchants and the slave trade, with people whose appearance was different from their own. By "many faces", however, we do not just mean people's typical appearance but also their character, how they behaved and how they were perceived by people outside their own sphere. The written sources paint one picture, the archaeological another, and with this study osteology makes its contribution. It can be said of the written sources that they depict the political situation and the affairs of different kin groups, the mythology and the way foreign groups viewed the Norse, while archaeological findings show us everyday customs associated with people both high and low on the social scale. Archaeology contributes palpable evidence about material culture, with both indigenous and foreign objects testifying how people dressed and adorned themselves. It also shows how great the variation was between people in the same community and between people in different regions.

In the mortuary practice we find a wide range of customs, which in turn reflects the different groups in society. The Viking Age, as we have seen, was a time of change when new customs were mixed with old. The practice of cremating the dead was abandoned and inhumation was introduced. The changes were partly due to the influence of Christianity, and presumably the change occurred even before Christianity was fully adopted. Inhumation occurred before people began to be buried in the Christian-style east–west direction. For a long time both pagan and Christian burial customs existed side by side. We can thus say that people at this time displayed at least three different faces in their mortuary practice: those who stuck to the old beliefs and customs, those who were in between, and those who left the old ways behind and adopted the new. All three can be found at the originally pagan burial grounds. We can thus envisage that there were burials of different religious character at the same place, perhaps even simultaneously. Even during pagan times our Viking Age burial grounds show variation in the treatment of the dead, but this disappeared with Christianity when everyone had to be treated in the same way. In this study, the graveyard Trinitatis in Skåne clearly manifest this change. In terms of wealth, as judged by the number of artefacts in the graves, the burial grounds at Birka are the most original with objects revealing characteristics of a trading site with wide networks, but also in magnitude compared with the burial sites in Vannhög and Fjälkinge in Skåne.

Speaking of faces, it is not every face one wants to see at a burial. In many cemeteries in Scandinavia, there are bodies which were buried on their bellies, often with the face towards the ground, turned away from the people taking part in the funeral. The reasons why some people were buried in this way are unknown, but there are several suggested explanations. One assumption is that it is a sign of exclusion from the community. Perhaps it was someone who, at a certain stage in life, had somehow turned his back on the others and was therefore buried with his face down as a sign of humiliation or punishment. Gotland was distinguished in this context and the question is why? Analysis of strontium isotopes and artefacts in the graves did not indicate that this group was more likely to embrace non-locals. Most of them are found at the coastal burial sites and especially at Kopparsvik. Possibly it reflects the special nature of the place, a collection of people without family nearby, someone who could be responsible for the treatment of the dead. Those who exhibit strontium values typical of Gotland can originally have their homestead several kilometres from the site. This, in turn, indicates that relatives were not aware of the death until the person in question was already buried. The person was not transported home to the local grave field. Among other things, this phenomenon, along with several others, suggests that the graveyard was probably a burial place for people who were in the vicinity for a short time.

The Viking Age is best known for people's increasing mobility, often with distant travels, but some people also moved around within present-day Sweden or Scandinavia. Earlier research claimed that the reason for the increased mobility was overpopulation. Now scholars believe instead that the Norse expansion and raiding were triggered by a desire for riches, not so much that people emigrated out of necessity. Another factor suggesting that people did not leave because of any difficulty in providing for themselves is that people on average were relatively tall. Stature is a parameter that, even in today's society, indicates prosperity.

Through analysis of strontium isotopes, we have been able to ascertain how many of the individuals buried in one place had their roots there and how many came from elsewhere. The results vary from one burial place to another, but we can say without doubt that a very large proportion are not local. A clear pattern has also emerged, namely, that in some places there are representatives of a group that may have its origin among the western Slavs. This group is represented at all three burial grounds in Skåne, on Öland and Gotland, but there are no examples as know of in the Mälaren region. In Lund it looks as if these people did not come here voluntarily. They are all buried in the periphery of the graveyard. Perhaps these are the potters responsible for the manufacture of the Baltic ware, made of domestic clay but with Slavic inspiration. This group may, for several reasons, have displayed a different kind of face, reflected in a different cultural origin and perhaps even a different appearance. Another pattern is seen in the indications on Öland and Gotland that people from the Mälaren valley either settled on the island or died while visiting it and were therefore buried there. The former is perhaps more probable. People from the Mälaren area do not appear to have moved as far south as the Danish kingdom. Perhaps the results are an indication of the political situation.

We also conclude that in Trinitatis, Fjälkinge, Öland and Birka, the proportion of non-locals is higher than, for example, at Vannhög, Gotland, Galgedil and Trelleborg in Denmark. This could reflect the degree of networking. If we look more closely at specific individuals, we can see evidence that some brought their own customs with them. We may have found an example at the Kopparsvik cemetery on Gotland where a man was buried on a bearskin. Bear does not occur naturally on Gotland and the man did not come from the island. Perhaps the custom where he came from was that bodies were buried on an animal skin, whether sheep, goat, or a wild animal such as a bear (Fig. 50). He was rather young and may have been a hunter or a fur merchant. Another example is an elderly woman who had 14 Arabian silver coins scattered on her body. The strontium value shows that she was not originally from the island. Perhaps this was the custom where she came from. We should also point out, however, that she, like yet another woman who was not from the island, had the typical Gotlandic animal-head brooches in her grave; they were probably attached to the clothes in which she was buried. Another interesting finding is that, while previous research has indicated that those who practised cremation in Birka were local, this study demonstrates that, of the five who were analyzed, only one came from Björkö, and there were also individuals who did not even have their roots in the Mälaren valley.

Regarding the question as to what extent the non-locals seem to have been an integrated part of the community, the results show that it varies from place to place. At Trinitatis in Lund, several of the non-locals have retained their way of burying, which may be because they belonged to a group who did not come there voluntarily, and they have not, to the same extent, adapted the local customs. For example, they used their own types of coffins made of old trough. On Gotland, on the other hand, we see that non-locals also used the typical Gotlandic animal-shaped buckles. We have also not found that non-locals were buried with a different set of objects than the locals. Divergent positions in the grave, for example on the stomach, were found among both locals and non-locals.

Migration also means that illnesses of various kinds can be spread with individuals travelling or migrating to the Nordic countries, and one of the diseases that began to spread more rapidly during the Viking Age, affecting Scandinavia, was leprosy. Although a case that is several hundred years earlier has been noted in Halland, there are no finds of leprosy showing any continuity up to the Viking Age. The individual, a woman, may have come from outside Halland. At any rate, she was treated differently when she was buried, probably because of her changed appearance. She is the only one among 150 burials who was not cremated.[1] In the mound where her remains were placed there was already a cremated individual, and later the remains of another cremated person were deposited on top of the leprous woman. As the disease develops, people with leprosy change their appearance completely. In our study the strontium analyses show that several of the individuals with the disease did not come from the places where they were buried, and some even came from outside Scandinavia. Most cases of leprosy found hitherto have been in present-day Skåne, so it is not strange that the first leper hospital was established there too. Mobility at the end of the Viking Age no doubt contributed to the relatively quick spread of the disease in the 11th century.

The Viking Age is known for being a violent period. The osteological results show surprisingly few traces of violence, whether healed or not. Was the strategy that they used when raiding to attack people who were unprepared and therefore unarmed? Or is the reason that the Vikings were on the warpath more in the early Viking Age, when people in Scandinavia mostly cremated their dead, with the result that we cannot identify traces of healed injuries? Is there anything to indicate that slaves were treated badly, that they have more healed fractures than other people? The great problem here is the difficulty of identifying individuals who belonged to the group of slaves. In Lund there is one possible case. It is a woman who lived sometime between 990 and 1050/60 and whose strontium value indicates that she was not local. She suffered from leprosy and had died from a blow to her forehead. Her placing at the edge of the graveyard together with other bodies with low strontium values, along with the fact that she had leprosy, may suggest that she was one of the people brought to Lund from Slavic territory to serve as labour. Perhaps she was one of the makers of the Baltic ware that occurs so frequently in the town.

The osteological studies have shown that people in the Viking Age suffered several of the diseases that are still found in contemporary society, just as the face of a person with stiff joints today can show grimaces and tired features when aches and pains make it difficult to perform the simplest tasks such as washing and dressing. One person who suffered greatly was the woman with rheumatoid arthritis who lived and died at Fjälkinge. The accepting attitude of Viking Age society is indicated by the study of individuals who were born with disabilities. The fact that they reached a great age and were buried like other people suggests that they were well looked after from the beginning and were included in the community. Examples of this are seen in the dwarfs who have been found. Dwarfism is extremely rare in archaeological skeletal material simply because it is a rare condition. We know of only three cases in Scandinavia, and all three are from the Viking Age; moreover, two of them were found in the same burial place in Skämsta because it is such a rare disease, they must have been related. It was not easy to cope with a child who had that condition; it shows that they grew up in a society with an understanding for abnormalities. In Sweden today there are 20–30 people with dwarfism caused by this disease.

In today's society the classical signs of old age appear later and later in life. One such sign is dental health. Fifty years ago it was not unusual to see people with sunken gums after they had placed their dentures in a glass of water for the night. Today modern dental care has improved artificial teeth so that they do not need to be taken out, and facial features are thus the same day and night. Many young people in the Viking Age had good teeth, but caries was also a frequent occurrence. Jaw inflammations resulted in toothache and swollen cheeks, and old people lost many of their teeth, their lips became wrinkled, and the mouth sunk in.

Many archaeological excavations of burial grounds from prehistoric societies, including the Viking Age, show that it was not uncommon for the smallest children to be few in number, if there are any at all. The explanation proposed for this is that not all children were buried and that people in the Viking Age practised infanticide. Such things no doubt happened, but it seems unlikely that this was so common that it affected the demographic picture on the scale indicated by the findings of many excavations. This study comprising several different burial grounds suggests other interpretations, for example, that some burial places could have been divided according to age. Certain cemeteries may also have served groups who were temporarily staying at the place. In that respect our study helps to nuance the picture of how infants were perceived, showing that they were considered worthy of a burial.

Finally, we have one of the most striking facial expressions investigated in this study, the male custom of filing horizontal grooves in the teeth. This was a distinct modification of the face which lasted for life, and which also gives us an opportunity to study bodily decoration. Some written sources hint that the Vikings painted their bodies, but because skin is not preserved from the Viking Age, the filed grooves on the teeth are the only clear evidence of bodily modification. A global comparison indicates that filed teeth were probably intended to mark that a man belonged to a group of some kind. Besides giving Vikings a face that was specific for the time, it signalled that certain individuals wanted to stand out in this way. The comparison also shows that the custom was practised elsewhere in Europe. Finds both in graves and at settlement sites from the Viking Age have demonstrated contacts with people from far afield, and therefore we feel bold enough to suggest that there may be a link to the isolated find of horizontally filed teeth in Egypt. The Norsemen need not have been in Egypt, but both parties may have met at a common marketplace on one or more occasions and derived the inspiration there.

With this study we hope that we have pointed to possible general conclusions about life in the Viking Age by examining burial grounds from several different geographical areas. Osteology not only contributes knowledge about the majority of the population, those who were not buried with the finest grave goods or with the most splendid monuments, but also adds to our knowledge about life at the time. Through the study we want to increase interest in large-scale demographic investigations. We anticipate that the results of the study, in combination with the improved opportunities afforded by aDNA, will lead to further knowledge about population size, and give some idea of how large the immigrant element was and where these people came from. The strontium analyses have also give us good insight into the extent of mobility, and we hope that this method will be used more often on both old and new excavations. By showing what the combination of several disciplines can achieve, we hope that a study like this can inspire continued cooperation across subject boundaries, and also that scientific methods will be described and used in a way that can benefit many.

Appendix: Strontium values

Table 1 Strontium values from animals, used for baseline.

Site	Co-ordinates	Landscape	Species	Strontium value	Reference	Map legend Fig 44a
Rya	(56.230503, 13.162474)	S	pig	0.712	A	1
Rya	(56.230503, 13.162474)	S	pig	0.712	A	1
Uppåkra	(55.666753, 13.171284)	S	cattle	0.711	A	A
Uppåkra	(55.666753, 13.171284)	S	cattle	0.711	A	A
Uppåkra	(55.666753, 13.171284)	S	cattle	0.711	A	A
Uppåkra	(55.666753, 13.171284)	S	cattle	0.711	A	A
Uppåkra	(55.666753, 13.171284)	S	cattle	0.712	A	A
Uppåkra	(55.666753, 13.171284)	S	cattle	0.712	A	A
Uppåkra	(55.666753, 13.171284)	S	cattle	0.712	A	A
Uppåkra	(55.666753, 13.171284)	S	cattle	0.712	A	A
Uppåkra	(55.666753, 13.171284)	S	cattle	0.712	A	A
Uppåkra	(55.666753, 13.171284)	S	cattle	0.712	A	A
Uppåkra	(55.666753, 13.171284)	S	cattle	0.713	A	A
Uppåkra	(55.666753, 13.171284)	S	cattle	0.713	A	A
Uppåkra	(55.666753, 13.171284)	S	cattle	0.712	A	A
Uppåkra	(55.666753, 13.171284)	S	cattle	0.712	A	A
Uppåkra	(55.666753, 13.171284)	S	cattle	0.712	A	A
Uppåkra	(55.666753, 13.171284)	S	pig	0.711	A	A
Uppåkra	(55.666753, 13.171284)	S	pig	0.711	A	A
Uppåkra	(55.666753, 13.171284)	S	pig	0.711	A	A
Uppåkra	(55.666753, 13.171284)	S	pig	0.711	A	A
Uppåkra	(55.666753, 13.171284)	S	pig	0.712	A	A
Uppåkra	(55.666753, 13.171284)	S	pig	0.712	A	A
Uppåkra	(55.666753, 13.171284)	S	water vole	0.711	A	A
Uppåkra	(55.666753, 13.171284)	S	water vole	0.711	A	A
Kvarteret Blekhagen	(5.704660, 13.191007)	S	black rat	0.711	A	A
Kvarteret Blekhagen	(5.704660, 13.191007)	S	black rat	0.711	A	A
Lindängelund	(55.557008, 13.010735)	S	water vole	0.711	A	A
Lindängelund	(55.557008, 13.010735)	S	pig	0.711	A	A
Trelleborg	(55.386632, 13.144327)	S	dog	0.710	B	A

Appendix: Strontium values

Table 1

Site	Co-ordinates	Landscape	Species	Strontium value	Reference	Map legend Fig 44a
Trelleborg	(55.386632, 13.144327)	S	dog	0.711	B	A
Jäsrrestad	(55.537520, 14.286635)	S	pig	0.713	A	2
Järrestad	(55.537520, 14.286635)	S	pig	0.713	A	2
Gårdstaliden	(56.103997, 13.713803)	S	roe deer	0.718	A	4
Gårdstaliden	(56.103997, 13.713803)	S	pig	0.716	A	4
Mölleholmen	(55.915497, 14.287267)	S	beaver	0.711	B	B
Transval	(55.922500, 14.283214)	S	fox	0.715	B	B
Hammar:1	(56.031688, 14.207799)	S	water wole	0.712	A	B
Hammar:2	(56.031688, 14.207799)	S	pig	0.712	A	B
Ringsjöholm	(55.900940, 13.394752)	S	water vole	0.711	A	3
Ringsjöholm	(55.900940, 13.394752)	S	water vole	0.711	A	3
Norje Sunnansund	(56.125048, 14.669034)	B	vole	0.717	C	C
Norje Sunnansund	(56.125048, 14.669034)	B	vole	0.717	C	C
Norje Sunnansund	(56.125048, 14.669034)	B	vole	0.718	C	C
Norje Sunnansund	(56.125048, 14.669034)	B	snail	0.719	C	C
Agerum	(56.145380, 14.638809)	B	snail	0.709	C	5
Kalmar	(56.663445, 16.356779)	SM	pig	0.717	D	E
Kalmar	(56.663445, 16.356779)	SM	cat	0.713	D	E
Kalmar	(56.663445, 16.356779)	SM	cat	0.713	D	E
Kalmar	(56.663445, 16.356779)	SM	pig	0.717	D	E
Näsby, Hultaby	(57.420997, 15.04308)	SM	hare	0.718	M	7
Näsby, Hultaby	(57.420997, 15.04308)	SM	dog	0.720	M	7
Målilla, Hägeråkra	(57.389140, 15.816221)	SM	micromam.	0.719	M	6
Glömminge, Brostorp	(56.719548, 16.536921)	Ö	dog	0.712	E	D
Glömminge, Brostorp	(56.719548, 16.536921)	Ö	dog	0.716	E	D
Glömminge, Brostorp	(56.719548, 16.536921)	Ö	micromam.	0.714	E	D
Glömminge, Brostorp	(56.719548, 16.536921)	Ö	micromam.	0.714	E	D
Norra Möckleby Bröttorp	(56.661007, 16.681595)	Ö	dog	0.720	E	D
Norra Möckleby Bläsinge	(56.664810, 16.681356)	Ö	dog	0.710	E	D
Norra Möckleby Bläsinge	(56.661007, 16.681595)	Ö	sheep/goat	0.713	E	D
Böda, Kronoparken	(57.260547, 17.035103)	Ö	dog	0.721	E	D
Mörbylånga	(56.523757, 16.386916)	Ö	dog	0.715	E	D
Gärdslösa, Sörby Störlinge	(56.784803, 16.737653)	Ö	dog	0.711	E	D
Gärdslösa, Sörby Störlinge	(56.784803, 16.737653)	Ö	micromam.	0.717	E	D

Appendix: Strontium values 97

Site	Co-ordinates	Landscape	Species	Strontium value	Reference	Map legend Fig 44a
Gärdslösa, Sörby Störlinge	(56.784803, 16.737653)	Ö	micromam.	0.718	E	D
Gärdslösa, Sörby Störlinge	(56.784803, 16.737653)	Ö	micromam.	0.712	E	D
Gärdslösa, Sörby Störlinge	(56.784803, 16.737653)	Ö	pig	0.719	E	D
Gärdslösa, Sörby Störlinge	(56.784803, 16.737653)	Ö	sheepgoat	0.712	E	D
Bredsättra, Ormöga	(56.859705, 16.780794)	Ö	micromam.	0.716	E	D
Bredsättra, Ormöga	(56.859705, 16.780794)	Ö	micromam.	0.712	E	D
Bredsättra, Ormöga	(56.859705, 16.780794)	Ö	micromam.	0.713	E	D
Bredsättra, Ormöga	(56.859705, 16.780794)	Ö	pig	0.712	E	D
Sandbyborg	(56.552624, 16.639111)	Ö	sheep/goat	0.714	E	D
Sandbyborg	(56.552624, 16.639111)	Ö	sheep/goat	0.712	E	D
Havdhem	(57.160728, 18.333269)	GO	fox	0.711	B	F
Vaskinde	(57.691496, 18.421393)	GO	hedgehog	0.710	B	F
Visby kungsladugård	(57.634800, 18.29484)	GO	hare	0.713	B	F
Visby	(57.634800, 18.29484)	GO	hare	0.710	B	F
Visby	(57.634800, 18.29484)	GO	hare	0.710	B	F
Visby	(57.634800, 18.29484)	GO	hare	0.711	B	F
Visby	(57.634800, 18.29484)	GO	plant	0.712	L	F
Visby	(57.634800, 18.29484)	GO	plant	0.713	L	F
Visby	(57.634800, 18.29484)	GO	plant	0.711	L	F
Visby	(57.634800, 18.29484)	GO	water	0.711	L	F
Fröjel	(57.335465, 18.190956)	GO	hare	0.710	B	F
Fröjel	(57.335465, 18.190956)	GO	cattle	0.713	F	F
Fröjel	(57.335465, 18.190956)	GO	pig	0.717	F	F
Fröjel	(57.335465, 18.190956)	GO	cattle	0.711	F	F
Fröjel	(57.335465, 18.190956)	GO	sheep	0.713	F	F
Fröjel	(57.335465, 18.190956)	GO	pig	0.733	F	F
Fröjel	(57.335465, 18.190956)	GO	cattle	0.712	F	F
Fröjel	(57.335465, 18.190956)	GO	cattle	0.715	F	F
Fröjel	(57.335465, 18.190956)	GO	pig	0.711	F	F
Fröjel	(57.335465, 18.190956)	GO	pig	0.711	F	F
Fröjel	(57.335465, 18.190956)	GO	pig	0.711	F	F
Fröjel	(57.335465, 18.190956)	GO	sheep	0.713	F	F
Fröjel	(57.335465, 18.190956)	GO	sheep	0.714	F	F
Fröjel	(57.335465, 18.190956)	GO	sheep	0.713	F	F
Gärdstad	(58.450000, 15.633333)	ÖG	sheep/goat	0.741	G	9
Gärdstad	(58.450000, 15.633333)	ÖG	pig	0.733	G	9
Gävbo, Östorp	(58.312382, 15.397581)	ÖG	micromam.	0.720	G	8
Kanaljorden	(58.538033, 15.047094)	ÖG	elk	0.721	H	G
Kanaljorden	(58.538033, 15.047094)	ÖG	wild boar	0.721	H	G

Appendix: Strontium values

Table 1

Site	Co-ordinates	Landscape	Species	Strontium value	Reference	Map legend Fig 44a
Kanaljorden	(58.538033, 15.047094)	ÖG	wild boar	0.726	H	G
Kanaljorden	(58.538033, 15.047094)	ÖG	wild boar	0.726	H	G
Kanaljorden	(58.538033, 15.047094)	ÖG	brow bear	0.739	H	G
Strandvägen	(58.537108, 15.062333)	ÖG	elk	0.735	H	G
Strandvägen	(58.537108, 15.062333)	ÖG	dog	0.725	H	G
Strandvägen	(58.537108, 15.062333)	ÖG	dog	0.725	H	G
Strandvägen	(58.537108, 15.062333)	ÖG	dog	0.721	H	G
Strandvägen	(58.537108, 15.062333)	ÖG	dog	0.712	H	G
Strandvägen	(58.537108, 15.062333)	ÖG	dog	0.725	H	G
Strandvägen	(58.537108, 15.062333)	ÖG	roe deer	0.734	H	G
Strandvägen	(58.537108, 15.062333)	ÖG	beaver	0.754	H	G
Strandvägen	(58.537108, 15.062333)	ÖG	beaver	0.728	H	G
Strandvägen	(58.537108, 15.062333)	ÖG	beaver	0.725	H	G
Strandvägen	(58.537108, 15.062333)	ÖG	red deer	0.731	H	G
Strandvägen	(58.537108, 15.062333)	ÖG	red deer	0.728	H	G
Strandvägen	(58.537108, 15.062333)	ÖG	red deer	0.731	H	G
Strandvägen	(58.537108, 15.062333)	ÖG	red deer	0.731	H	G
Strandvägen	(58.537108, 15.062333)	ÖG	red deer	0.733	H	G
Strandvägen	(58.537108, 15.062333)	ÖG	hedgehog	0.720	H	G
Strandvägen	(58.537108, 15.062333)	ÖG	hedgehog	0.720	H	G
Strandvägen	(58.537108, 15.062333)	ÖG	otter	0.724	H	G
Strandvägen	(58.537108, 15.062333)	ÖG	wild boar	0.721	H	G
Strandvägen	(58.537108, 15.062333)	ÖG	wild boar	0.726	H	G
Strandvägen	(58.537108, 15.062333)	ÖG	wild boar	0.726	H	G
Strandvägen	(58.537108, 15.062333)	ÖG	wild boar	0.724	H	G
Strandvägen	(58.537108, 15.062333)	ÖG	wild boar	0.722	H	G
Strandvägen	(58.537108, 15.062333)	ÖG	wild boar	0.723	H	G
Strandvägen	(58.537108, 15.062333)	ÖG	wild boar	0.733	H	G
Strandvägen	(58.537108, 15.062333)	ÖG	wild boar	0.726	H	G
Strandvägen	(58.537108, 15.062333)	ÖG	wild boar	0.721	H	G
Strandvägen	(58.537108, 15.062333)	ÖG	wild boar	0.723	H	G
Strandvägen	(58.537108, 15.062333)	ÖG	brown bear	0.740	H	G
Strandvägen	(58.537108, 15.062333)	ÖG	brown bear	0.740	H	G
Hjelmars rör	(58.175029, 13.553217)	VÄG	hare	0.716	I	H
Hjelmars rör	(58.175029, 13.553217)	VÄG	fox	0.714	I	H
Hjelmars rör	(58.175029, 13.553217)	VÄG	fox	0.714	I	H
Frälsegården	(58.460162, 13.853089)	VÄG	rodent	0.714	I	H
Frälsegården	(58.460162, 13.853089)	VÄG	rodent	0.715	I	H
Frälsegården	(58.460162, 13.853089)	VÄG	rodent	0.716	I	H
Frälsegården	(58.460162, 13.853089)	VÄG	rodent	0.714	I	H
Frälsegården	(58.460162, 13.853089)	VÄG	hedgehog	0.713	I	H
Frälsegården	(58.460162, 13.853089)	VÄG	rodent	0.714	I	H

Appendix: Strontium values

Site	Co-ordinates	Landscape	Species	Strontium value	Reference	Map legend Fig 44a
Rössberga	(58.139054, 13.336669)	VÄG	rodent	0.715	I	H
Rössberga	(58.139054, 13.336669)	VÄG	rodent	0.714	I	H
Rössberga	(58.139054, 13.336669)	VÄG	wildcat	0.724	I	H
Karleby Godegården	(58.175029, 13.553217)	VÄG	snail	0.712	I	H
Karleby Godegården	(58.175029, 13.553217)	VÄG	snail	0.714	I	H
Karleby Godegården	(58.175029, 13.553217)	VÄG	snail	0.714	I	H
Aranäs	(58.665735, 13.583878)	VÄG	pike	0.714	I	H
Aranäs	(58.665735, 13.583878)	VÄG	pike	0.725	I	I
Aranäs	(58.665735, 13.583878)	VÄG	elk	0.725	I	I
Aranäs	(58.665735, 13.583878)	VÄG	elk	0.725	I	I
Aranäs	(58.665735, 13.583878)	VÄG	red deer	0.726	I	I
Aranäs	(58.665735, 13.583878)	VÄG	mouse	0.726	I	I
Skara	(58.386013, 13.439328)	VÄG	pikeperch	0.726	I	12
Skara	(58.386013, 13.439328)	VÄG	pike	0.729	I	12
Skara	(58.386013, 13.439328)	VÄG	roach	0.727	I	12
Torpa	(57.649982, 13.279352)	VÄG	rodent	0.718	I	10
Torpa	(57.649982, 13.279352)	VÄG	rodent	0.719	I	10
Torpa	(57.649982, 13.279352)	VÄG	roe deer	0.718	I	10
Vädersholm	(57.839930, 13.282309)	VÄG	pike	0.715	I	11
Valbo, Lund	(60.643031, 17.008251)	GÄ	sheep/goat	0.764	J	15
Valbo, Hemlingby	(60.650540, 17.163769)	GÄ	dog	0.732	J	16
Stavby, Jönninge	(60.010236, 17.965346)	U	dog	0.73	J	13
Skuttunge, Eke	(59.998611, 17.514722)	U	dog	0.723	J	14
Fresta, Grimsta	(59.497055, 17.943926)	U	sheep/goat	0.731	J	K
Alsike, Åhusby	(59.749170, 17.77939)	U	rodents	0.725	J	K
Norrsunda	(59.272548, 18.740263)	U	rodents	0.726	J	K
Birka	(59.335216, 17.544778)	U	rodents	0.725	J	K
Birka	(59.335216, 17.544778)	U	rodents	0.727	J	K
Birka	(59.335216, 17.544778)	U	rodents	0.727	J	K
Birka	(59.335216, 17.544778)	U	rodents	0.728	J	K
Salberget, Sala	(59.920859, 16.606328)	V	sheep/goat	0.722	K	J
Salberget, Sala	(59.920859, 16.606328)	V	cow	0.737	K	J
Salberget, Sala	(59.920859, 16.606328)	V	chicken	0.737	K	J
Salberget, Sala	(59.920859, 16.606328)	V	pike	0.728	K	J
Salberget, Sala	(59.920859, 16.606328)	V	hare	0.737	K	J
Salberget, Sala	(59.920859, 16.606328)	V	pig	0.739	K	J

Key:
Landscape: Skåne = S; Småland = SM; Öland = Ö; Gotland = GO; Östergötland = ÖG; Västergötland = VÄG; Uppland = U; Västmanland = V
References: A = Price 2013; B = Price in this study; C = Kjällqvist & Price, Sunnansund and Norjegravfältet manuscripts; D = Magnell 2015; E = Wilhelmsson & Price 2017; F = Peschel 2014; G = Helander, Arcini and Evans manuscript; H = Eriksson, Frei, Howcroft, Gummesson, Molin, Lidén, Frei, Hallgren 2015; I = Sjögren, Price, Ahlström 2009; J = Price, Arcini,Drenzel, Gustin, Kalmring 2018; K = Bäckström & Price 2016; L) Frei unpublished; M Wilhelmsson and Price unpublished

Table 2 Human strontium values.

Place	Landscape	Grave no.	Riocarbon/*dendro-date (AD)	Age group	Sex	Stature	Burial area	Strontium value	Local	Non-local	Leprosy	Filed teeth
TR	S	1082		A	W	168	A	0.711	W			
TR	S	1103		A	M	189	A	0.710		M		
TR	S	1386		A	M	183	A	0.713		M		
TR	S	1450		A	M	177	A	0.713		M		
TR	S	1504		A	W	163	A	0.718		W		
TR	S	1512		A	W	152	A	0.711	W			
TR	S	104		A	W	163	B	0.709		W	x	
TR	S	112	1000±5 yrs	M	M	168	B	0.710		M		
TR	S	178		A	W	152	B	0.709		W		
TR	S	188		S	W	163	B	0.712	W			
TR	S	192		J	W	160	B	0.708		W	x	
TR	S	204	1054–1060	A	W	171	B	0.714		W	x	
TR	S	239	1032?	A	W	156	B	0.709		W		
TR	S	252		M	W	151	B	0.709		W		
TR	S	26	1017±5 yrs	M	M	162	B	0.710		M		
TR	S	27		S	W	170	B	0.714		W		
TR	S	272	1015±5 yrs	A	M	169	B	0.709		M		
TR	S	280		A	W	170	B	0.709		W		
TR	S	282		A	M	161	B	0.710		M		
TR	S	29	970–1160	A	M	176	B	0.710		M		x
TR	S	299		J/A	W	162	B	0.712	W			
TR	S	322		M	M	161	B	0.711	M			
TR	S	366		A	M	184	B	0.711	M			
TR	S	93	1013±5 yrs	M	W	158	B	0.717		W		
KO	GO	61		A	M	164	N	0.711	M			
KO	GO	58		A	M	179	N	0.710	M			
KO	GO	154		A	M	185	N	0.715		M		x
KO	GO	244		A	M	166	S	0.713	M			x
KO	GO	24	690–890	A	M	164	N	0.712	M			x
KO	GO	81		A	M	172	N	0.711	M			x
KO	GO	136		A	M	167	N	0.713	M			x
KO	GO	157		M	M	173	N	0.713	M			x
KO	GO	163	770–970	M	M	172	N	0.713	M			x
KO	GO	188		A	M	168	N	0.716		M		x

Appendix: Strontium values 101

Place	Landscape	Grave no.	Riocarbon/*dendro-date (AD)	Age group	Sex	Stature	Burial area	Strontium value	Local	Non-local	Leprosy	Filed teeth
KO	GO	72		A	M		N	0.716		M		
KO	GO	246		M	M	171	S	0.734		M		
KO	GO	324		J	W		?	0.714		W		
KO	GO	226		M	W	162	S	0.739		W	x	
KO	GO	141		S	W		N	0.713	W			
KO	GO	173		M	M	172	N	0.711	M			x
KO	GO	251		A	M		S	0.709		M		
KO	GO	164		A	W	163	N	0.709		W		
KO	GO	52		A	M	169	N	0.713	M			
KO	GO	15		A	M		N	0.713	M			x
KO	GO	20		J	M		N	0.714		M		
KO	GO	8		A	W		N	0.710	W			
KO	GO	3		A	M		N	0.714	M			
KO	GO	189		A	W	164	N	0.713	W			
KO	GO	280		M	M	180	S	0.714		M		x
KO	GO	151		M	M		N	0.710	M			
KO	GO	135		A	M	172	N	0.712	M			
KO	GO	235		A	M		S	0.709		M		x
KO	GO	14	890–1030	M	W	170	N	0.711	W			
KO	GO	260		M	M		S	0.711	M			
KO	GO	286		A	W		S	0.709		W		
KO	GO	140		M	W	149	N	0.711	W			
KO	GO	290		M	W	163	S	0.714		W		
KO	GO	277		A	W		S	0.712	W			
KO	GO	11		M	M		N	0.709		M		
KO	GO	22		A	W		N	0.714		W		x
KO	GO	112		S	W	164	N	0.717		W		
KO	GO	2		A	M		N	0.712	M			
KO	GO	270		A	W		S	0.710	W			
KO	GO	96		A	W	164	N	0.711	W			
KO	GO	291		S	W		S	0.710	W			
KO	GO	132		M	M	175	N	0.715		M		x
KO	GO	294		A	W		S	0.712	W			
KO	GO	283		M	M		S	0.710	M			
KO	GO	18		A	W		N	0.711	W			
KO	GO	230		A	M		S	0.714		M		
KO	GO	245		S	W	160	S	0.710	W			
SL	GO	2B/47	820–1000	A	W	154		0.711	W			

Table 2

Place	Landscape	Grave no.	Riocarbon/*dendro-date (AD)	Age group	Sex	Stature	Burial area	Strontium value	Local	Non-local	Leprosy	Filed teeth
SL	GO	24(A)		A	M	170		0.711	M			x
SL	GO	SK?			?			0.717		?		
SL	GO	2a/47		A	W	161		0.715		W		
SL	GO	2?		A	M			0.710	M			x
SL	GO	5			?			0.712	?			x
SL	GO	4		A	M			0.711	M			x
SL	GO	8	1040–1220	S	M	181		0.711	M			x
SL	GO	1/47		A	M	181		0.711	M			
SL	GO	5		A	M			0.709		M		
VA	S	1		J	?			0.710	?			
VA	S	2	650–870	A	M	177		0.710	M			
VA	S	4	660–900	M	M	183		0.713		M		x
VA	S	5		J	W?			0.709		W?		
VA	S	6		A	W			0.709		W		
VA	S	7	770–1020	M	M			0.710	M			
VA	S	8		M	M	165		0.709		M		
VA	S	12/13		M	W	158		0.711	W			
VA	S	14		A	?			0.711	?			
VA	S	200		A	M			0.709		M		
VA	S	307	680–970	A	M	170		0.710	M			
VA	S	373		A	?			0.719		?		
VA	S	400		A	M			0.710	M			
VA	S	598		A	M	170		0.710	M			
VA	S	745		A	M?			0.710	M			
VA	S	769		A	W			0.709		W		
VA	S	F11	690–980	A	M			0.711	M			x
FJ	S	1		A	M	172		0.710		M		
FJ	S	13		A	M	170		0.713		M	x	
FJ	S	49	720–950	M	M	162		0.710		M		x
FJ	S	56		A	W	166		0.714		W		
FJ	S	65		S	M	176		0.708		M		
FJ	S	77	780–980	J	M?			0.713		M	x	
FJ	S	86		A	M	174		0.710		M		
FJ	S	99		M	W	170		0.720		W	x	
FJ	S	223		S	W	158		0.715		W		
FJ	S	225	890–1030	A	W	158		0.710		W		
FJ	S	230		J	M			0.711	M			
FJ	S	304B		J	M?			0.711	M			

Place	Landscape	Grave no.	Riocarbon/*dendro-date (AD)	Age group	Sex	Stature	Burial area	Strontium value	Local	Non-local	Leprosy	Filed teeth
FJ	S	363		M	M	161		0.709		M		
FJ	S	366		S	W	161		0.710		W		
FJ	S	387		A	M	185		0.711	M			
FJ	S	428A		S	W	162		0.712	W			
FJ	S	430		A	M	160		0.710		M		
FJ	S	446 IV		M	W	171		0.710		W		
FJ	S	545		M	M	168		0.709		M	x?	
FJ	S	600		A	M?	168		0.710		M		
FJ	S	658		A	M	160		0.710		M		
FJ	S	729B		A	M	178		0.713		M	x?	
FJ	S	744		M	W	162		0.710		W		
FJ	S	771		S	W	153		0.711	W			
FJ	S	772		A	M	168		0.712	W			
FJ	S	866		M	M	175		0.710		M		
FJ	S	954		J	M	166		0.709		M		
FJ	S	1013		A	M	169		0.710		M		
FJ	S	1020B		M	M	167		0.710		M		
FJ	S	1129		A	M	171		0.711	M			
FJ	S	349		S	W	169		0.709		W		
FR	GO	20/01		A	W?		N	0.715		K?		
FR	GO	00702		A	M		N	0.714		M		
FR	GO	33/04B		A	W	160	N	0.709		W		
FR	GO	352004		A	M	170	N	0.711	W			
FR	GO	00902		A	M		S	0.714		M		
FR	GO	1601		A	W	164	N	0.713	W			
FR	GO	21a/00		A	W	164	N	0.715		W		
FR	GO	38/04b		A	W	168	N	0.709		W		
FR	GO	38/04a		A	W	163	N	0.716		W		
FR	GO	36/01		A	M	175	N	0.710	M			
FR	GO	8b/04		A	W	163	N	0.709		W		
FR	GO	25a/00		A	W	158	N	0.711	W			
FR	GO	32/88		A	W	155	N	0.709		W		
FR	GO	12/88		A	W	155	N	0.717		W		
FR	GO	01100		A	W		N	0.710	W			
FR	GO	5/03		A	W?		N	0.711	W			
FR	GO	00303		A	W		N	0.714		W		
FR	GO	3		A	W		N	0.710	W			
FR	GO	1		A	M?		N	0.714		M		
FR	GO	2/87		A	W		N	0.712	W			

Table 2

Place	Landscape	Grave no.	Riocarbon/*dendro-date (AD)	Age group	Sex	Stature	Burial area	Strontium value	Local	Non-local	Leprosy	Filed teeth
FR	GO	03604		J	M	166	N	0.712	M			
FR	GO	22/00		A	W		N	0.716		W		
FR	GO	2303		A	W		N	0.715		W		
FR	GO	19a/89		A	M	167	N	0.712	M			
FR	GO	3a/88		A	W	159	N	0.715		W		
FR	GO	1/1987		A	m	174	N	0.710	M			
FR	GO	2b/1990		J	?		N	0.712	?			
FR	GO	1/89		A	W?		N	0.710	M			
FR	GO	2/89		?	?		N	0.713	?			
FR	GO	046a-99		M	M		N	0.716		M		
FR	GO	37		A	M		N	0.727		M		
FR	GO	19b/89		J	M?		N	0.715		M		
FR	GO	24/89		J/A	M		N	0.711	M			
FR	GO	32-99		A	k		N	0.721		W		
FR	GO	500		?	?		N	0.715		?		
FR	GO	1a/90		A	M?		S	0.711	M			
FR	GO	31990		A	?		N	0.712	?			
FR	GO	4/90		A	?		N	0.712	?			
FR	GO	9/88		A	M	176	N	0.715		M		
FR	GO	600		A	W		N	0.709		W		
FR	GO	22		A	M		N	0.711	M			
FR	GO	39		A	M		N	0.735		M		
FR	GO	25b/89		A	W		N	0.715		W		
FR	GO	2301		J	W		N	0.711	?			
FR	GO	1701		A	M?		N	0.724		M		
FR	GO	15/99		A	M		N	0.709		M		
FR	GO	11/89		A	W?		N	0.713	W			
FR	GO	34/01		A	?		N	0.708		?		
FR	GO	016/99		A	W		N	0.711	W			
FR	GO	02404		A	W	155	N	0.710	W			
FR	GO	41-99		A	W		N	0.710	W			
FR	GO	1202		A	M?		S	0.715		M		
FR	GO	9/89		A	W		N	0.713	W			
FR	GO	42		A	?		N	0.712	?			
FR	GO	28		A	M		N	0.718		M		
FR	GO	1/90B		A	M		S	0.712	M			
FR	GO	032-99E		?	?		N	0.708		?		
FR	GO	33-04A		A	W		N	0.709		W		

Place	Landscape	Grave no.	Riocarbon/*dendro-date (AD)	Age group	Sex	Stature	Burial area	Strontium value	Local	Non-local	Leprosy	Filed teeth
FR	GO	02800		A	W		N	0.713	W			
FR	GO	48		A	W		N	0.716		W		
BI	U	114	cremation	-	?			0.72				
BI	U	141	cremation	-	?			0.719				
BI	U	275		A	?			0.716				
BI	U	496		A	M			0.734				
BI	U	512b		J/A	?			0.718				
BI	U	557:I		A	W?			0.727	W?			
BI	U	557:II		A	W			0.718		W		
BI	U	566:01		A	M			0.71		M		
BI	U	566:02		A	M			0.716		M		
BI	U	566:03		A	W			0.717		W		
BI	U	605b		A	?			0.715		?		
BI	U	585		A	W			0.711		W		
BI	U	607:II		A	?			0.731		?		
BI	U	620		A	M			0.721		M		
BI	U	638		A	?			0.723		?		
BI	U	642		Inf II	?			0.718		?		
BI	U	643		A	?			0.728	?			
BI	U	644:I		J	M			0.718		M		
BI	U	644:2		A	M			0.725	X			
BI	U	770		Inf II	?			0.723		?		
BI	U	793		Inf II	?			0.728	X			
BI	U	804		A	?			0.719		?		
BI	U	834		A	?			0.729		?		
BI	U	841		A	?			0.717		?		
BI	U	855		A	M			0.734		M		
BI	U	860b		A	?			0.73		?		
BI	U	865		A	M			0.732		M		
BI	U	869	cremation	-	?			0.714		?		
BI	U	930	cremation	-	?			0.72		?		
BI	U	946		A	?			0.719		?		
BI	U	950		Inf II	?			0.717		?		
BI	U	954		M	M			0.729		M		
BI	U	962		J/A	?			0.733		?		
BI	U	967		A	W			0.727	W			
BI	U	1012:II		A	M			0.711		M		
BI	U	1012:I		A	?			0.713		?		
BI	U	1015	cremation	-	?			0.724		?		

Table 2

Place	Landscape	Grave no.	Riocarbon/*dendro-date (AD)	Age group	Sex	Stature	Burial area	Strontium value	Local	Non-local	Leprosy	Filed teeth
BI	U	1030		A	M			0.719		M		
BI	U	1036		Inf I	?			0.718		?		
BI	U	1053		A	?			0.731		?		
BI	U	1097		A	W			0.713		W		
BI	U	1997		A	M			0.713		M		

Key:
Place: Tr = Trinitatis; Lund, Fr = Fröjel; Bi = Birka, Björkö; Ko = Kopparsvik; Sl = Slite; Fj = Fjälkinge; Va = Vannhög
Landscape: Sk = Skåne; GO = Gotland; UP = Uppland
Age group: Inf I = 1–6 yrs; Inf II = 7–14 yrs; J = 15–19 yrs; A = 20–39 yrs; M = 40–59 yrs; S = 60+
Sex: W = Woman; M = Men; ? = Not determined
Burial area: A = near the church; B = outer part of cemetery; N = north; S = south

Table 3 Men with filed teeth and their strontium value.

Strontium value	Trinitatis	Vannhög	Fjälkinge	Öland	Gotland	Birka	Total
0.707							
0.708				1 NL			1
0.709					1 NL		1
0.710	1 Nl		1 NL		1 L	1 NL	4 PG
0.711		1 L			7 L		8 PG
0.712					2 L		2 PG
0.713		1 NL		1 L	4 L		6 PG
0.714					2 NL		2
0.715					1 NL		1
0.716					1 NL		1
0.717							
0.718							
0.719							
0.720							
0.721							
0.722							
0.723							
0.724							
0.725							
0.726							
0.727							

Table 3

Strontium value	Trinitatis	Vannhög	Fjälkinge	Öland	Gotland	Birka	Total
0.728							
0.729							
0.730							
0.731							
0.732							
0.733							
0.734						1 NL	1
0.735							
0.736							
0.737							
0.738							
0.739							

L = Local, NL = Non local, Potentially from Gotland = PG

Notes

Chapter 1

1. Svanberg 2016
2. Andersson 2016; Svanberg 2016
3. Svanberg 2016
4. Bennike 1985; 1994; Holck 2006; 2009; Price *et al.* 2010; 2015; Kjellström 2012; Naumann *et al.* 2014a; 2014b; Wilhelmsson and Ahlström 2015; Wilhelmsson and Price 2017
5. Bennike 1994
6. Kjellström 2012
7. Kjellström 2005
8. Holck 2006
9. Nauman *et al.* 2014a
10. Wilhemson and Ahlström 2015; Wilhelmson and Price 2017
11. Wilhelmson and Ahlström 2015; Wilhelmson and Price 2017
12. Price *et al.* 2012; 2015
13. Price *et al.* 2010
14. Buikstra and Übelaker 1994
15. Sjøvold 1990

Chapter 2

1. Gräslund 1980
2. Jacobsen and Moltke 1942
3. Cinthio 2018
4. Carelli 2004
5. Cinthio 2002
6. Cinthio 2002
7. Cinthio 2002
8. Andrén 2000
9. Cinthio 2002
10. Cinthio 2002
11. Roslund 2001
12. Cinthio 2002
13. Cinthio 2002
14. Madsen 1990
15. Carelli 2003
16. Cinthio 2002
17. Arcini 1999
18. Blomqvist 1941; 1951; Blomqvist and Mårtensson 1963; Cinthio 1992; 1996; 2002
19. Arcini 1999
20. Jacobsson 1995
21. Jacobsson 1995
22. Klint Jensen 1973
23. Christiansen 1971
24. Roesdahl 1988
25. Jacobsson 1995
26. Jacobsson 1995
27. Hansson 1993; Becker 2001 and Hellerström 2005 and references therein
28. Arcini and Jacobsson 2008
29. Helgesson 1993
30. Helgesson 1990; 1993
31. Helgesson 1990
32. Helgesson 1990
33. Arcini 1991
34. Thunmark-Nylén 2004
35. Toplak 2016
36. Thunmark-Nylén 2006; Toplak 2016
37. Thunmark-Nylén 2006
38. Toplak 2016
39. Thunmark-Nylén 2006 and references therein
40. Thunmark-Nylén 2000a; 2000b; 2006
41. Toplak 2016
42. Larje 1987
43. Petterson 1966
44. Sörling 1939; 1945
45. Einerstam and Thålin 1946 and Lindström and Schützler 1974 and references therein
46. Mortágua 2005–2006
47. Carlsson 1999
48. Carlsson 1999
49. Carlsson 1999
50. Vos 2005 and references therein
51. Stolpe 1872; 1876; 1878; 1879a; 1879b; 1882
52. Ambrosiani 1997
53. Holmquist Olausson 2001
54. Arbman 1943
55. Ambrosiani 2012
56. Ambrosiani 2012
57. Ambrosiani 2002; 2005; Ambrosiani and Gustin 2002; Arbman 1939; 1943; Hedenstierna-Jonson 2001; Jansson 1997

58 http://www.birkavikingastaden.se/2016/04/birkaportalen/
59 LBI, and Gunnar Andersson pers. comm.
60 Gräslund 1980
61 Arbman 1943
62 Gräslund 1980
63 Holmquist Olausson 1993; 1997; Andersson 2016
64 Cleve 1948; Lamm 1973a; 1973b
65 Gräslund 1980
66 Stolpe 1889
67 Gräslund 1980
68 Andersson 2010
69 Ambrosiani 2016
70 http://www.birkavikingastaden.se/2016/04/birkaportalen
71 Kjellström 2016
72 Gräslund 1980
73 Kjellström 2016
74 Gräslund 1980
75 Frölund 2000
76 Frölund 2000
77 Frölund and Larsson 1995
78 Arcini 1995
79 Stolpe 1882; Almgren 1904; Stenberger 1964; Gräslund 1980; Cinthio 2002; Price 2002; Kjellström et al. 2005; Svanberg 2003; Helgesson and Arcini 1996; Thunmark-Nylén 1995; 2006; Carlsson 1999; Andersson 1999
80 Arbman 1943; Gräslund 2010
81 Arcini 2009
82 Arcini 2009 and reference therein
83 Halsall 2002
84 Arcini 2009 and reference therein; Mielella et al. 2015
85 Toplak 2016

Chapter 3

1 Roesdahl 1988; Sawyer 1999; Hjardar and Vike 2016; Hedenstierna-Jonson 2015; Price 2002
2 Vedeler 2014
3 Sawyer 1999; Hjardar and Vike 2016
4 Sawyer 1999
5 Sawyer 1999
6 Roesdahl, 1988; Sawyer 1999
7 Sawyer 1999
8 Adam of Bremen 1984; Brink 2012; Roslund 2006
9 Price 1985; Sillen and Kavanagh 1982
10 Price et al. 2010; 2015; Wilhelmson and Ahlström 2015; Wilhelmson and Price 2017
11 Pechel 2014
12 Wilhelmson and Ahlström 2015; Wilhelmson and Price 2017
13 Frei and Price 2012
14 Arcini et al. 2014
15 Price 2013
16 McGovern et al. 2006
17 Price et al. 2018
18 Larsson 2013
19 Blomqvist and Mårtenson 1963; Roslund 2006; Cinthio 2018
20 Roslund 2001; 2006
21 Douglas Price personal communication
22 Frei and Price 2012

23 Evans et al. 2010
24 Cinthio 2018
25 Arcini et al. 2014
26 Brorsson 2003; Roslund 2006
27 Edring 1997
28 Paananen-Kemi 2011; Roslund 2001
29 Roslund 2001
30 Roslund 2006 and Roslund, in press.
31 Price et al. 2018
32 Cinthio 2018
33 Lundström 1976; Cinthio 2002
34 Petré and Wigardt 1973
35 Petré and Wigardt 1973
36 Petré 1980
37 Lindsted 1997
38 Carlsson 1983
39 Segerstedt 1907; Jonsson 1993
40 Lindström and Schützler 1974
41 Arbman 1943
42 Hasselberg 1944; Nevéus 1974; Lindkvist 1979; Karras 1988
43 Hemmendorf 1984; Zachrisson 1994; Skre 1998; Brink 2012; Zachrisson 2014
44 Hemmendorf 1984; Holmquist 1979
45 Naumann et al. 2014
46 Nevéus 1974
47 Iversen 2003
48 Hasselberg 1944; Nevéus 1974
49 Zachrisson 2014
50 Funegård Viberg 2012

Chapter 4

1 WHO 1946; Naido and Willis 2009
2 Sellevold et al 1984; Bennike 1994; Arcini 1999; Arcini et al. 2014
3 Buko 2014; Steckel 1995
4 Arcini et al. 2014
5 Steckel 1995; Koepke and Baten 2005; Hatton 2014
6 Hatton 2014
7 Tanner et al. 1982; Bielicki et al. 1986
8 Kuh et al. 1997; Meyer and Selmer 1999
9 Tanner et al. 1982; Bielicki et al. 1986
10 Welinder et al. 1998
11 Magnell 2017
12 Scott and Julie 2008
13 Bennike 1994
14 Lindahl et al. 1995; Artelius 1989
15 Heimdahl 2009
16 Bennike and Brade 1999
17 Lawrence 1977
18 Arcini 1992
19 Birkner 1978; Resnick and Niwayama 1989; Lemley 2009
20 Arcini 1991
21 Arcini 1999
22 Tracy et al. 2004
23 Hjardar and Vike 2016
24 Braathen 1989
25 Peets et al. 2010

26 Enoksen 2004
27 Hjardar and Vike 2016
28 Hjardar and Vike 2016
29 Hjardar and Vike 2016
30 Arcini 1999
31 Ällmäe *et al.* 2011
32 Price *et al.* 2016
33 Loe *et al.* 2014
34 Arcini 1999; 2003; Menander and Arcini 2012.
35 http://www.socialstyrelsen.se/ovanligadiagnoser/medfoddspondyloepifysealdyspla
36 Arcini and Frölund 1996
37 Larje 1985
38 http://www.socialstyrelsen.se/ovanligadiagnoser/medfoddspondyloepifysealdyspla
39 Arcini 1995
40 Bryceson and Pfaltzgraff 1979
41 Arcini and Artelius 1993
42 Blondiaux *et al.* 2002
43 Inskip *et al.* 2015
44 Dzierzykray-Rogalski 1980
45 Robbins *et al.* 2009
46 Chakrabarty and Dastidar 1989
47 Hansen 1874
48 Bryceson and Pfaltzgraff 1979
49 Bryceson and Pfaltzgraff 1979
50 Arcini 1999
51 Economou *et al.* 2013
52 Blomqvist 1946

Chapter 5

1 Rastorfer 2004
2 Insoll 2015
3 Samadelli *et al.* 2015
4 Spindler 2000
5 Rubin 1988
6 Ko 1994; 1997
7 Kenny and Nichols 2017
8 Meiklejohn *et al.* 1992; Tiesler 1999
9 Steele 2001
10 Steele 2001
11 von Jhering 1882; Romero 1970; De La Borbolla 1940; Fastlich 1962; Milner and Larsen 1991; Mower 1999; Tiesler *et al.* 2005; Arcini 2005; Bennike 2010; Alt and Pichler 1998; Übelaker 1987; Vukovic *et al.* 2009. Zumbroich and Salvador-Amores 2009; Afsin *et al.* 2013; Burnett and Irish 2017
12 Finucane *et al.* 2008; Zumbroich and Salvado Amores 2010; Schroeder and Haviser 2012
13 Romero 1958; Becker 1973; Smith 1972
14 Fastlicht 1962; Willey *et al.* 1965; Saul and Hammond 1973; Saul and Frank 1997; Havill *et al.* 1997; Massey and Steele 1997; Tiesler Blos 2001
15 Zumbroich 2011
16 Zumbroich 2011
17 Williams and White 2006
18 Kennedy *et al.* 1981
19 Finucane *et al.* 2008

20 Arcini 2005
21 Mortágua 2006
22 Arcini 2005
23 Mortágua 2006; Kjellström 2014
24 Arcini 2005
25 Vretemark, pers. comm.
26 Kjellström 2014; Arcini and Drenzel this study
27 Kjellström 2014
28 Kjellström 2014; Hennius *et al.* 2016
29 Sjøvold, Drenzel, and Arcini this study
30 Arcini, Drenzel, and Karlsson this study
31 Arcini 2005
32 Kjellström 2014
33 Lundgren 1975
34 Arbman 1943
35 Hägg 1983; 2003
36 Arcini 2010
37 Kjellström 2014
38 Hemmendorf 1984
39 Wilhelmson and Price 2017
40 Smith 1996
41 Bennike 2010
42 Loe *et al.* 2014
43 Loe *et al.* 2014
44 Loe *et al.* 2014
45 Stewart and Titterington 1944; 1946; Milner 1983
46 Farjas unpublished
47 Agusti Farjas *et al.* in press
48 Buzon *et al.* 2007
49 Van Reenen 1986; Reichart *et al.* 2008; Schroeder and Haviser 2012
50 The last Journals of David Livingstone, in Central Africa. From eighteen hundred and sixty-five to his death. Continued by a narrative of his last moments and sufferings. Obtained from his faithful Servants, Chuma and Susi, by Horace Waller. F.R.G.S., rector of Twywell, Northhampton 1875-01-01
51 Insoll 2015
52 Vedeler 2014
53 Price *et al.* 2016
54 Buko 2014
55 Wikander 1978
56 Adam of Bremen 1984
57 Hägg 2003
58 Linderholm *et al.* 2008
59 Lundgren 1975
60 Androshchuk 2008
61 Östergren 2003

Chapter 6

1 Arcini *et al.* 2006; 2014
2 Fiorato *et al.* 2000; Syse 2003; Loe *et al.* 2014
3 Mogren 1984
4 Auler 2012; Karlsson 2008
5 Frölund 2000; Arcini 2008
6 Nagmér 1983
7 Classon 2004; Mejsholm 2009
8 Pentikäinen 1968, 75; Clover 1988; Mundal 1989

9   Pentikäinen 1968; 1990; Gräslund 1983
10 Cinthio 2018
11 Nilsson 2012
12 Mejsholm 2009; Arcini 1999
13 Gejvall 1960
14 Cinthio 1999
15 Helgesson, pers. comm.
16 Nelsson 1987
17 Pettersson 2006
18 Pettersson 2006; Ingvarsson-Sundström 2005
19 Schulze 1984
20 Strömberg 1950; Nagmér 1983
21 Nagmér 1983
22 Westholm 1988
23 Yrving 1992
24 Jansson 1983; Yrving 1992
25 Thunmark-Nylén 2004
26 Östergren 1983
27 Toplak 2016
28 Thunmark-Nylén 2004
29 Bodin 1987
30 Björk 2013
31 Lindström and Schützler 1974
32 Gräslund 1980
33 Toplak 2016

Chapter 7

1   Arcini and Artelius 1993

References

Adam of Bremen. 1984. *Historien om Hamburgerstiftet och dess biskopar*, Trans. into Swedish by E. Svenberg. Profide et Christianissimo 6(184). Prorius. Stockholm.

Afsin, H., Sadi Cagdir, A. Büyük, Y. and Karaday, B. 2013. Cosmetic dentistry in ancient times: V-shaped dental mutilation in skeletal remains from Corycus, *Bulletin of the International Association for Paleodontology* 7(2): 148–156.

Agusti, B., Pons, E. and Arcini, C. in press. Marcas de desgaste dentario ritual en el período bizantino de Oxirrinc (Minia, Egipto). In Díaz Zorita, M. and López Flores, I. (eds), *XIII Congreso Nacional de Paleopatología. Paleopatología y bioarqueología. Contextualizando el registro óseo. Écija 2015*.

Agustí Farjas, B. in press., Informe d'antropologia. In Padró Parcerissa, J. *Memòria provisional de les excavacions de la campanya de 2014 a El-Bahnasa, Oxirrinc (Mínia, Egipte)*. Nilus (Aegiptolgy Catalan Association magazine), Barcelona n. 23.

Allmäe, R., Maldre, L. and Tomek, T. 2011. The Salme I ship burial: an osteological view of a unique burial in Northern Europe. *Interdisciplinaria Archaeologica, Natural Sciences in Archaeology* 2011(2): 109–24.

Almgren, O. 1904. *Vikingatidn grafskick i verkligheten och i den fornnordiska literaturen*, 309–46. Uppsala.

Alt, K. W. and Pichler, S. L. 1998. Artificial modifications of human teeth. In Alt, K. W., Rösing, F. W. and Teschler-Nicola, M. (eds), *Dental Anthropology: Fundamentals, Limits, and Prospects*, 387–415. New York.

Ambrosiani, B. 1997. Birka – stad i nätverk. In Ellerströ, J. (ed.), *Amico amici*, 111–17. Lund.

Ambrosiani, B. 2002. *Birka: the site and its eastern connections*. Stockholm.

Ambrosiani, B. 2005 Birka and Scandinavia's trade with the east. In Kovalev, R. And Sherma, H. M. (eds), *Festschrift for Thomas S. Noonan* 2, 287–96. Idyllwild.

Ambrosiani, B. 2012. Birkaforskning i perspektiv av grävningarna 1990–1995. In Hedenstierna-Jonsson, C. (ed.), *Birka. Nu*, 9–18. Stockholm.

Ambrosiani, B. 2016. *Hantverkare och handel i Birka*. Stockholm.

Ambrosiani, B. and Gustin, I. 2002. Islamic links to Scandinavia during the Viking Age. In *Velikij volžskij put': materialy Meždunarodnoj naučno-praktičeskoj konferenczii "Velikij volžskij put', Kazan'- Astrachan'- Kaza*n*'*, 6–16 avgusta 2001 g. 2, Istorija formirovanja i razvitija, 253–5. Kazan.

Andersson, G. 1999. *Gravspråk som religiös strategi. Valsta och Skälby Attunaland under vikingatid och tidig medeltid*. Riksantikvarieämbetet arkeologiska undersökningar, skrifter 61. Stockholm.

Andersson, G. 2016. Meet the Vikings. In Andersson, G. (ed.), *We Call Them Vikings*, 9–15. Stockholm.

Andersson, K. 2010. *Glas från romare till vikingar*. Uppsala.

Andrén, A. 2000 Ad sanctos – de dödas plats under medeltiden. Middelalderens kirkegårde. *Hikuin* 27, 7–26.

Androshchuk, F. 2008. *Vikings in the East: essays on contacts along the road to Byzantium (800–1100)*. Uppsala.

Arbman, H. 1939. *Birka: Sveriges äldsta handelsstad*. Stockholm.

Arbman, H. 1943. *Birka: Untersuchungen und Studien. 1, Die Gräber*. Kungl. Vitterhets historie och antikvitets akademien, Stockholm.

Arcini, C. 1991. *Vikingatida Fjälkingbor: ståtliga vikingar i minoritet*. Kristianstad.

Arcini, C. 1992. Rheumatoid arthritis – rare reality as recovered among Scanian skeletal remains from Viking and medieval times. *Sydsvenska Medicinhistoriska*. 18: 11–21.

Arcini, C. 1995. Osteologisk undersökning av gravarna från RAÄ 203, Uppland, Tierp socken, Skämsta. RAÄ dnr 2750/93UV Syd Osteologisk Rapport 1995: 5.

Arcini, C. 1999. Health and Disease in early Lund. *Health and disease in early Lund: osteo-pathologic studies of 3,305 individuals buried in the first cemetery area of Lund 990–1536*. Acta Archaeologica Lundensia. Lund.

Arcini, C. 2003. *Åderförkalkning och portvinstår. Välfärdssjukdomar i medeltidens Åhus. Riksantikvarieämbetet*, Stockholm.

Arcini, C. 2005. The Vikings bare their filed teeth. *American Journal of Physical Anthropology* 128(4): 727–33.

Arcini, C. 2008. Detta lämnar ingen oberörd. In Fendin, T., *Döden som straff*, 68–103. Linköping.

Arcini, C. 2009. Loosing face. The worldwide phenomenon of ancient prone burial. In Back, I.-M., Danielsson, I. Gustin, A. Larsson, N. Myrberg and S. Thedeén (eds), *Döda personers sällskap. Gravmaterialens identiteter och kulturella uttryck/On the Threshold. Burial Archaeology in the Twenty-First Century*, 187–202. Stockholm.

Arcini, C. 2010. Kopparsvik. Ett märkligt gravfält från vikingatid. *Gotländskt arkiv. Meddelanden från föreningen Gotlands Fornvänner* 82, 11–20.

Arcini, C. and Artelius, T. 1993. Äldsta fallet av spetälska i Norden. Lepra fanns redan i yngre romersk järnålder. *Arkeologi i Sverige:* 55–71.

Arcini. C. and Frölund, P. 1996. Two dwarves from Sweden: a unique case. *International Journal of Osteoarchaeology* 6: 155–66.

Arcini, C. and Jacobsson, B. 2008. Vikingarna från Vannhög. *Ale* 1: 1–14.

Arcini, C., Jacobsson, B. and Helgesson, B. 1991. Vikingatida hantverkare med slitspår på tänderna. *Populär arkeologi* 9: 19–21.

Arcini, C. Jacobson, B. and Persson, B. 2006. *Pestbacken.* Stockholm.

Arcini, C., Price, T. D., Cinthio, M., Drenzel. L., Andersson, M., Persson, B., Menander, H., Vretemark, M., Kjellström, A., Hedvall, R. and Tagessson, G. 2014. Living conditions in time of plague. In Lagerås, P. (ed.), *Environment, Society and the Black Death: an interdisciplinary approach to the late-medieval crisis in Sweden*, 104–40. Oxford.

Artelius, T. 1989. Boplatslämning vid Skottorps Säteri. In Artelius, T. and Lundqvist, L. (eds), *Bebyggelse – kronologi: boplatser från perioden 1800 f. Kr- 500 e. Kr i södra Halland. Nya bidrag till Halland äldsta historia* 2. Stockholm.

Auler, J. 2012. *Richtstättenarchäologie* 3. Dormagen.

Bäckström, Y. and Price, T. D. 2016. Social identity and mobility at a pre-industrial mining complex, Sweden. *Journal of Archaeological Science* 66: 154–68.

Becker, M. J. 1973 Archaeological evidence for occupational specialization among the Classic Period Maya at Tikal, Guatemala. *American Antiquity* 38: 396–406.

Becker, N. 2001. Kvarteret Verkstaden. Sönderplöjda gravar från yngre järnålder. Förundersökning inom RAÄ 2, Kkv. Verkstaden 3 i Trelleborg, Skåne. Arkeologisk förunderökning. *Riksantikvarieämbetet UV Syd Rapport 2001:16.*

Bennike, P. 1985. *Paleopathology of Danish Skeletons. A Comparative Study of Demography, Disease and Injury.* Copenhagen.

Bennike, P. 1994. An anthropological study of the skeletal remains of Vikings from Langeland. In Grøn, O., Hedeager Krag, A. and Bennike, P. (eds) *Vikingetidsgravpladser på Langeland*, 168–83. Rudkøbing.

Bennike, P. 2010. Tandtatovering, *Skalk*: 3–8. Højbjerg.

Bennike, P. and Brade, A-L. 1999. *Middelalderens sygdomme og behandlingsformer i Danmark.* Medicinsk-Historisk Museum København Universitet. Copenhagen.

Bielicki, T., Waliszko, Q., Hulanicka, B. and Kotlarz, K. 1986. Social-class gradients in menacheal age in Poland. *Annals of Human Biology.* 13: 1–11.

Birkner, R. 1978. *Normal Radiologic Patterns and Variances of the Human Skeleton: an X-ray atlas of adults and children.* Baltimore.

Björk, N. 2013. *Värdig ett vapen. En analys och tolkning av Birkas vapengravars gravgåvor och kontext.* Kandidatuppsats VT 2013. Högskolan Gotland.

Blomqvist, R. 1941. *Tusentalets Lund.* Lund.

Blomqvist, R. 1946. Sankt Jörgens hospital i Lund. *Humanitet* 1946(12), 208–12.

Blomqvist, R. 1951. *Lunds Historia 1. Medeltiden.* Lund.

Blomqvist, R. and Mårtensson, A. W. 1963. *Thulegrävningen 1961: en berättelse om vad grävningarna för Thulehuset i Lund avslöjade.* Acta Archaeologica Lundensia. Lund.

Blondiaux, J., Dürr, J., Khouchaf, L. and Eisenberg, L. E. 2002. Microscopic study and X-ray analysis of two 5th century cases of leprosy: paleoepidemiological inferences. In Roberts, C.A., Lewis, M. E. and Manchester, K. (eds), *The Past and Present of Leprosy. Archaeological, historical, paleopathological and clinical approaches. Proceedings of the International Congress on the Evolution and Paleoepidemiology of the Infectious Diseases* 3 (ICEPID). British Archaeological Report S1054. Oxford.

Bodin, U. 1987. *Vapengravar i Mälarområdet. En studie av vapenfrekvens och vapenkombinationer under folkvandringstid, vendeltid och vikingatid.* Uppsala. https://www.diva-portal.org/smash/get/diva2:632827/FULLTEXT01.pdf

Braathen, H. 1989: *Ryttergraver. Politiske strukturer i eldre rikssamlingstid.* Universitetets Oldsaksamling Varia 19. Oslo.

Brink, S. 2012. *Vikingarnas slavar: den nordiska trälddomen under yngre järnålder och äldsta medeltid.* Stockholm.

Brorsson, T. 2003. The Slavonic Feldberg and Fresendorf pottery in Scania, Sweden. In Larsson, L. and Hardh, B. (eds), Centrality – regionality: the social structure of southern Sweden during the Iron Age, 223–34. Acta Archaeologica Lundensia 40. Lund.

Bryceson, A. and Pfaltzgraff, R. E. 1979. *Leprosy.* Edinburgh.

Buikstra, J. E., and Ubelaker, D. 1994. *Standards for Data Collection from Human Skeletal Remains.* Arkansas Archeological Survey Research Series 44. Feyetteville.

Buko, A. 2014. *Bodzia: a late Viking-age elite cemetery in Central Poland. East Central and Eastern Europe in the Middle Ages, 450–1450.* Leiden.

Burnett, S. E. and Irish, J. D. 2017. *A World View of Bioculturally Modified Teeth (Bioarchaeological Interpretations of the Human Past.* Project MUSE. Gainesville.

Buzon, M. R., Simonetti, A. and Creaser, R. 2007. Migration in the Nile Valley during the New Kingdom period: a preliminar strontium isotope study. *Journal of Archaeological Science* 34, 1391–401.

Carelli, P. 2005. *Från Vespasianus till Leopold II: en sammanställning av jordfunna mynt i Lund 1833–1994.* Acta Archaeologica Lundensia. Lund.

Carelli, P. 2004. Lunds äldsta kyrkogård och förekomsten av ett senvikingatida danskt parochialsystem. In Lund, N. (ed.), *Kristendommen i Danmark før 1050*, 253–8. Roskilde.

Carlsson, A. 1983. *Djurhuvudformiga spännen och gotländsk vikingatid.* Stockholm Studies in Archaeology. Stockholm.

Carlsson, D. 1999. *Gård, hamn och kyrka. En vikingatida kyrkogård i Fröjel.* Visby.

Chakrabarty, A. N. and Dastidar, S.G. 1989. Correlation between occurrence of leprosy and fossil Fuels: role of fossil fuel bacteria in the origin and global epidemiology of leprosy. *Indian Journal of Experimenta Biology* 27: 483–96.

Christiansen, T. E. 1971. Traeningslejr eller tvangsborg. *Årbog for jysk arkeologisk selskab, KUML* 1970, 43–64. Århus.

Cinthio, M. 1992. Några daterings- och tolkningsproblem aktualiserade i samband med bearbetningen av gravar och kyrkogård tillhörande Trinitatiskyrkorna i Lund, *Meta* 1–2.

Cinthio, M. 1996. *Kyrkorna kring Kattesund.* Arkeologiska rapporter från Kulturen i Lund. Lund.

Cinthio, M. 1999. Trinitatiskyrkan i Lund/med engelsk prägel. Kirkearkeologi i Norden. *Hikuin* 24 (1997).

Cinthio, M. 2002. *De första stadsborna: Medeltida gravar och människor i Lund*. Eslöv.

Cinthio, M. 2018. Mission och etablering. In Cinthio, M. and Ödman, A. (eds), *Vägar mot Lund*, 11–120. Lund.

Classon, P. 2004. Den rituella handlingens praxis: exemplifierat av ett järnåldersgravfält beläget i mellersta Bohuslän. In Claesson, P. and Munkenberg B-A. (eds), *Gravar och ritualer*, 151–200. Uddevalla.

Cleve, N. 1948. Spår av tidig kristendom i västra Finland. *Finskt Museum* 1947–48: 65–85. Helsingfors.

Clover, C. 1988. The politics of scarcity: notes on the sex ratio in Early Scandinavia. *Scandinavian Studies* 60 (*Norse Values and Society*): 147–88.

De La Borbolla, R. 1940. Types of mutilation found in Mexico. *American Journal of Physical Anthropology* 26(1), 349–65.

Dzierzykray-Rogalski, T. 1980. Paleopathology of the Ptolematic inhabitants of Dakhleh Oasis. *Journal of Human Evolution* 9: 71–4.

Economou, C., Kjellström, A., Lidén, K., and Panagopoulos, I. 2013. Ancient DNA reveals an Asian type of Mycobacterium leprae in medieval Scandinavia. *Journal of Archaeological Science* 40: 465–70.

Edring, A. 1997. *Ett gårdskomplex från Vikingatid&tidig medeltid, Fjälkinge 48:15, Skåne. Rappport 9*. Kristianstad.

Einerstam, B. and Thålin, H. 1946. Torget i Slite – ett vikingatidsgravfält. *Gotländskt Arkiv* 101–11.

Enoksen, L. M. *Vikingarnas stridskonst*. Lund

Eriksson, G., Frei, K. M., Howcroft, R., Gummesson, S., Molin, F., Lidén, K., Frei, R. and Hallgren, F. 2015. Diet and mobility among Mesolithic hunter-gatherers in Motala (Sweden) – The isotope perspective. *Journal of Archaeological Science*: Reports https://www.sciencedirect.com/search?authors=eriksson&pub=Journal%20of%20Archaeological%20Science%3A%20Reports&show=25&sortBy=relevance

Evans, J. A. Montgomery, J., Wildman, G. and Boulton, N. 2010. Spatial variations in biosphere 87Sr/86Sr in Britain. *Journal of the Geological Society* 167, 1–4, 7 January 2010, https://doi.org/10.1144/0016-76492009-090

Fastlich S. 1962. Dental inlays and fillings among the ancient Mayas. *Journal of the History of Medicine and Allied Sciences* 17: 392–401.

Finucane, B. C., Manning, K. and Touré, M. 2008. Prehistoric dental modification in West Africa – early evidence from Karkarichinkat Nord, Mali. *International Journal of Osteoarchaeology* 18(6): 632–40.

Fiorato, V., Boylston, A. and Knüsel, C. 2000. *Blood Red Roses: the archaeology of a mass grave from the Battle of Towton AD 1461*. Oxford.

Forsander, K.V. 1931. Undersökningsrapport. *Antikvariskt topografiskt arkiv (ATA), Riksantivarieämbetet* dnr:0684/31.

Frei, K. M. and Price, T. D. 2012. Strontium isotopes and human mobility in prehistoric Denmark. *Archaeological Anthropological Science* 4: 103–14. DOI 10: 1007/s12520-011-0087-7.

Frölund, P. 2000. Död och begravd för 1000 år sedan. Människor i norra Uppland. Uppland 2000. *Årsbok för medlemmarna i Upplands fornminnesförening och hembygdsförbund*: 11–34.

Frölund, P. and Larsson, L-I. 1995. Skämsta – bosättning och gravar i norra Uppland. Arkeologisk undersökning Fornlämning 203 och 342 Skämsta 1:11, 1:14, 3:3, 3:5 Tierps socken Uppland. *Riksantikvarieämbetet UV Uppsala Rapport* 1997: 67.

Funegård Viberg, O. 2012. *Fängslande begravningar. En studie av påmage-gravar på gravfälten på Gotland under vikingatiden*. Kandidatuppsats i arkeologi. Stockholm.

Gejvall, N-G, 1960. *Westerhus: medieval population and church in the light of skeletal remains*. Lund.

Gräslund, A-S. 1980. *Birka IV, The Burial Customs. A Study of the graves on Björkö*. Motala.

Gräslund, A-S. 1983. The human sex ratio at birth. Reproductive strategies in early society. Annales Acaddemiae Regiae Scientarium Upsaliencsis Kungl. *Vetensapssamhällets i Uppsala Årsbok* 24 (1981–1982), 59–81. Uppsala.

Gräslund, A-S. 2010. Religionsskiftet speglat i gravskicket. Ny svensk forskning kring senvikingatida gravar och gravskick. In Nilsson, B. (ed.), *Från hedniskt till kristet. Förändringar i begravningsbruk och gravskick i Skandinavietn ca: 800–1200*, 141–64. Runica et Meiviala. Oputuscal 14. Stockholm.

Hägg, I. 1983. Birkas orientaliska praktplagg. *Fornvännen* 78: 204–23.

Hägg, I. 2003. Härskarsymbolik i Birkadräkten. In editors? *Dragt og magt: studier af magtsymboler i dragten*, 14–27. Copenhagen.

Halsall, P. 2002. *The Book of Sugar Abbot of St Deniseon What Was Done During his Administration (c. 1144–48)* http://www.fordham.edu/halall/source/sugar.htlm

Halvill, L. M., Warren, D. M., Jacobi, K. P., Gettelman, K. D., Cook, D. C. and Pyburn, K. A. 1997. Late Classic tooth filing at Chau Hiix and Tipu, Belize. In Wittington, S. L. and Reed, D. M. (eds) *Bone of Maya: Study of Ancient Skeletons*, 89–104. Washington DC.

Hansen, G. A. 1874. *Undersøgelser Angående Spedalskhedens Årsager*. Norwegian Master Degree in Medicine 4, 1–88. Oslo.

Hansson, M. 1993. Neolitiska boplatslämningar och vikingatida graver. Arkeologisk för och slutundersökning. 1990. Skåne, Trelleborgs sn. Västervång 2:97 (Kv Verkstaden) samt arkeologisk rapport. *Riksantikvarieämbetet UV Syd. ATA, Rapport dnr: 6344/90 och 6446/90*.

Hasselberg, G. 1944. Den s.k. Skarastadgan och träldomens upphörande i Sverige. *Västergötlands fornminnesförenings tidskrift* del 5, häfte 3.

Hatton, T. J. 2014. How have Europeans grown so tall? *Oxford Economic Papers Advance Access* 66(2): 349–72.

Hedenstierna-Jonson, C. 2001. Befästa handelsstäder, garnisoner och professionella krigare. In Olausson, M. (ed.), *Birkas Krigare*, 65–72. Stockholm.

Hedenstierna-Jonson, C. 2015. Close encounters with the Buzantine border zone: on the eastern connections of the Birka warrior. In Minaeva, O. and Holmquist, L. (eds), *Scandinavia and the Balkans: cultural interactions with Byzantium and Eastern Europe in the first millennium AD*, 139–52. Newcastle upon Tyne.

Heimdahl, J. 2009. Bolmörtens roll i magi och medicin under den svenska förhistorien och medeltid. *Fornvännen* 104: 112–28.

Helgesson, B. 1990. *Rapport. Arkeologisk undersökning 1990. Fjälkinge 35:60 m fl Fjälkinge socken. Fornlämning 18 och 19*. Kristianstad.

Helgesson, B. 1993. *Gravarna berättar. Fjälkinge för 1000 år sedan*. Kristianstad.

Helgesson, B. and Arcini, C. 1996. A major burial ground discovered at Fjälkinge. Reflections of life in a Scanian Viking village, *Lund Archaeological Review* 2, 51–61.

Hellerström, S. 2005. *Gravar från yngre järnålder. Skåne, Trelleborgs kommun, Kvarteret Vekstaden 3*. RAÄ 2. *Lund Arkeologisk slutundersökning. Riksantikvarieämbetet, UV Syd, Dokumenta-tion av fältarbetsfasen (DAFF)* 2005: 5.

Hemmendorf, O. 1984. Människooffer. Ett inslag i järnålderns gravritualer, belyst av ett fynd i Bollstanäs, Uppland. *Fornvännen* 79: 4–12.

Hennius, A., Sjöling, E. and Prata, S. 2016. Människor kring Gnistahögen. Begravningar från vendeltid, vikingatid och tidig medeltid. Danmark 62:1, 127:1 & 227 Danmarks socken Uppsala kommun Uppland, *SAU RAPPORT* 2016:10.

Hjardar, K. and Vike, V. 2016. *Vikings at War*. Havertown.

Holck, P. 2006. The Oseberg ship burial, Norway: New thoughts on the skeletons from the grave mound. *European Journal of Archaeology* 9(2–3): 185–210.

Holck, P. 2009. The skeleton from the Gokstad ship: new evaluation of an old find. *Norwegian Archaeological Review* 42(1), 40–9. doi: 10.1080/00293650902907706.

Holmquist, L. 1979. Älgmannen från Birka: presentation av en nyligen undersökt krigargrav med människooffer. *Fornvännen* 85: 175–82.

Holmquist Olausson, L. 1993. *Aspects on Birka. Investigations and Surveys 1976–1989*. Theses and papers in Archaeology B3. Stockholm.

Holmquist Olausson, L. 1997. Birlas Borg efter avslutad undersökning. In Åkerlund, A., Bergh, S., Nordbladh, J., and Taffinder, J. (eds), *Till Gunborg Arkeologiska samtal*, 399–405. Archaeological Report 33. Stockholm.

Holmquist Olausson, L. 2001. Birkas befästningsverk – resultat från de senaste årens utgrävningar. In Olausson, M. (ed.), *Birkas krigare*, 9–15. Stockholm.

Ingvarsson-Sundström, A. 2005. Osteologisk analys. Skelettgravar från Triberga. RAÄ 73, Hulterstad sn, Öland. SAU Rapport 2005:14 O. In Petersson, M., *Triberga kvarnbacke, ett gravfält från yngre järnålder Hulterstad sn, Öland*. Arkeologiska enheten, rapport 2006: Kalmar.

Inskip, S. A., Taylor, M., Zakrzewski, S. R., Mays, S. A., Pike, A. W. G., Llewellyn, G., Williams, C. M., Lee, O. Y-C. L., Wu, H. H. T., Minnikin, D. E., Besra, G. S. and Stewart, G. R. 2015. Osteological, biomolecular and geochemical examination of an early Anglo-Saxon case of lepromatous leprosy. *PLoS ONE* 10(5) http://dx.doi.org/10.1371/journal.pone.0124282.

Insoll, T. 2015. *Material Explorations in African Archaeology*. Oxford.

Iversen, T. 2003. Fra trell til leilending – Frigivning, bosetning og herredømme i Norge, Norden og Vest-Europa i tidlig middelalder. In Lindkvist, T. and Myrdal, J. (eds), *Trälar. Ofria i agrarsamhället från vikingatid till medeltid*, 22–50. Skrifter om skogs- och landbrukshistoria 17. Stockholm.

Jacobsen, L. and Moltke, E. 1942. *Danmarks runeindskrifter tex-atlas. København 1941–1942*, Copenhagen.

Jacobsson, B. 1995. *Trelleborgen – en av Harald Blåtands danska ringborgar*. Trelleborg.

Jansson, I. 1983. Gotland och omvärlden under vikingatiden – en översikt. In Jansson, I. (ed.), *Gutar och Vikingar*, 207–47. Stockholm.

Jansson, I. 1997. Warfare, trade or colonisation? Some general remarks on the eastern expansion of the Scandinavians in the viking period. In Hansson, P. (ed.), *The Rural Viking in Russia and Sweden*, 9–64. Örebro.

Jonsson, K. 1993. A Gotlandic hoard from the early Viking age. In Arwidsson, G. (ed.), *Sources and Resources: studies in honour of Birgit Arrhenius*, 451–8. Rixensart.

Karlsson E. 2008. Glömda gravar på galgbacken. In Fendin, T. (ed.), *Döden som straff*, 12–67.

Karras, R. M. 1988. *Slavery and Society in Medieval Scandinavia*. Yale Historical Publications 135. New Haven.

Kennedy, K. A. R., Misra, V. N. and Burrow, C. B. 1981. Dental mutilation in prehistoric India. *Current Anthropology* 22: 285–6.

Kenny, E. and Nichols, E. G. 2017. *Beauty around the World. A Cultural Encyclopedia*. Santa Barbara.

Kjellström, A. 2005. *The Urban Farmer: osteoarchaeological analysis of skeletons from medieval Sigtuna interpreted in a socioeconomic perspective*. Stockholm.

Kjellström, A. 2012. Projektet Människor i brytningstid. Skelettgravar i Birka och dess nära omland. In Hedenstierna-Jonsson, C. (ed.), *Birka. Nu*, 69–80. Stockholm.

Kjellström, A. 2014. Spatial and temporal trends in new cases of men with modified teeth from Sweden (AD 750–1100). *European Journal of Archaeology* 17(1): 45–59.

Kjellström, A. 2016. People in transition: life in the Mälaren Valley from an osteological perspective. In Turner, V. (ed.), *Shetland and the Viking World*, 197–202. Lerwick.

Kjellström, A., Tesch, S. and Wikström, A. 2005. Inhabitants of a sacred townscape. An archaeological and osteological analysis of skeletal remains from late Viking Age and medieval Sigtuna, Sweden. *Acta Archaeologica* 76(2): 87–110. DOI: 10.1111/j.1600-0390.2005.00036.x.

Klint Jensen, O. 1973. *The Problem of Evaluating Archaeological Sources in Early Historical Time*. Actes du VIIe Congres international des Sciences Préhistoriques. Beograd.

Ko, D. 1994. *Teachers of the Inner Chambers: Women and Culture in Seventeenth-century China*. Stanford.

Ko, D. 1997. The body as attire: the shifting meanings of foot binding in seventeenth-century China. *Journal of Women's History* 8(4): 8–27. doi:10.1353/jowh.2010.0171.

Koepke, N. and Baten, J. 2005. The biological standard of living in Europe during the last two millennia. *European Review of Economic History* 9(1), 61–95.

Kuh, D., Power, C. and Blane, D. 1997. Social pathways between childhood and adult health. In Kuh, D. and Ben-Sclomo, Y. (eds), *A Life Course Approach to Chronic Disease*, 169–200. Oxford.

Lamm, J.P. 1973a. *Fornfynd och Fornlämningar på Lovö. Arkeologiska studier kringen uppländsk järnåldersbygd*. Theses and papers in North-European Archaeology 3. Stockholm.

Lamm, J.P. 1973b. En folkvandringstida kammargrav vid Torsätra. *Fornvännnen* 68, 81–9.

Larje, R. 1985.The short Viking from Gotland. A case study. In Backe, M. (ed.), *Archaeology and Environment* 4, 259–71. Umeå.

Larje, R. 1987. Preliminär osteologisk könsbeämning. Kopparsvik, Visby. Unpublished report.

Larsson, L. 2013. Rich women and poor men: analysis of a cemetery at Önsvala in the hinterland of Uppåkra. In Hårdh, B. and Larsson, L. (eds), *Folk, fä och fynd*, 133–61. Acta Archaeologica Lundensia Series 64. Lund.

Lawrence, J.S. 1977. *Rheumatism in Populations. Consultant to the Arthritis and Rheumatism.* Manchester.

Lemley, K.V. 2009. Kidney disease in nail–patella syndrome. *Pediatric Nephrology* 24(12): 2345–54. doi: 10.1007/s00467-008-0836-8PMCID: PMC2770138.

Lindahl, J., Lagerås, P. and Regnell, M. 1995. A deposition of bark vessel, flax and opium poppy from 2500 B.P. Sallerup, southern Sweden. In Robertsson, A-M., Hackens, T., Hicks, S., Rirberg, J. and Åkerlund, A. (eds), *Landscapes and Life. Studies in Honour of Urve Miller* (PACT 50), 305–30. Rixenart.

Linderholm, A., Hedenstierna Jonson, C., Svensk, O. and Lidén, K. 2008. Diet and status in Birka: stable isotopes and grave goods compared. *Antiquity* 82, 446–61.

Lindkvist, T.H. 1979. Landborna i Norden under äldre medeltid. *Studia Historica Upsaliensia*, 110. Uppsala.

Lindstedt, J. 1997. *Vikingatida mynt i gravar.* Seminarieuppsats i arkeologi vid Stockholms Universitet. Unpublished.

Lindström, M. and Schützler, L. 1974. Två vikingatida gravfält på Gotland Slite torg, Othem sn Laxarve 1. Stockholms Universitet. Unpublished.

Livingston, D. 1874. The last Journals of David Livingstone, in Central Africa. From eighteen hundred and sixty-five to his death. Continued by a narrative of his last moments and sufferings. Obtained from his faithful Servants, Chuma and Susi, by Horace Waller. F.R.G.S., rector of Twywell, Northhampton 1875-01-01, London.

Loe, L. Boyle, A. Webb, H. and Score, D. 2014. *Given to the Ground. A Viking Age Mass Grave on Ridgeway Hill, Weymouth.* Dorset Natural History and Archaeological Society Monograph 22. Oxford.

Lundgren, M. 1975. Två typer av silverband från en ryttargrav på Gotland. *Fornvännen* 70, 144–6.

Lundström, S. 1976. Båtdetaljer. In Mårtensson, A.W. (ed.), *Uppgrävt förflutet för PKbanken i Lund. En investering i arkeologi*, 135–43. Lund.

Madsen, P.K. 1990. *Han ligger under en blå sten. Hikuin* 17. Højbjerg.

Magnell, O. 2015. Djurhållning, jakt, fågelfångst och fiske. In Stibeus, M., *Kalmar slott. Bebyggelse och fynd från 1100–1800-talen.* National Historical Museum, Reportt 2015, 54. Linköping.

Magnell, O. 2017. Gårdarnas djur – osteologisk analys av Ostkustbanan genom Gamla Uppsala Rapport 2017: 1-2. Arkeologisk undersökning Uppsala län; Uppsala kommun; Uppsala socken; Gamla Uppsala 20:1, 21:13, 21:27 m.fl. Uppsala 134:4, 240:1, 284:2, 586:1, 597:1, 603:1, 604:1, 606:1, och 682. Report, Stockholm.

Mälarstedt, H. 1979. Kopparsvik – ett vikingatida gravfält vid Visby. In Falck, W. (ed.), *Arkeologi på Gotland*, 99–104. Visby.

Massey, V. K., and Steele, D. G. 1997. A Maya skull pit from the Terminal Classic Period, Colha, Belize. In Whittington, S. L. and Reed, D. M. (eds), *Bones of the Maya: Studies of Ancient Skeletons*, 62–77. Washington DC.

McGovern, T. H., Perdikaris, S., Einarsson, A. and Sidell, J. 2006. Coastal connections, local fishing, and sustainable egg harvesting: patterns of Viking Age inland wild resource use in My´vatn district, Northern Iceland. *Environmental Archaeology* 11(2), 187–205.

Meiklejohn, C., Anagnostis, A., Akkermans, P. A., Smith, P. E. L. & Solecki, R. (1992) Artificial cranial deformation in the Proto-Neolithic and Neolithic Near East and its possible origin: evidence from four sites, *Paléorient* 18(2), 83–97.

Mejsholm, L. 2009. *Gränsland: konstruktion av tidig barndom och begravningsritual vid tiden för kristnandet i Skandinavien.* Diss. Occasional papers in archaeology, 1100-6358; 44. Uppsala.

Menander, H. and Arcini, C. 2012. *Gravar i S:t Olofs konvent.* Skänningen 2:1, 3:1, Skänningen stad, Mjölby kommun, Östergötland. Riksanatikvarieämbetet. UV rapport 2012, 47. Linköping.

Meyer, H. E. and Selmer, R. 1999. Income, educational level and body height. *Annals of Human Biology* 26(3), 219–23.

Milella, M., Mariotti, V., Belcastro, M. G. and Knüsel, C. J., 2015. Patterns of Irregular Burials in Western Europe. PLoS ONE 10(6):e0130616. doi:10.1371/journal.pone.0130616.

Milner, G. R. 1983. *The East St. Louis Stone Quarry Site Cemetery.* Urbana.

Milner, G. R. and Larsen, C-S. 1991. Teeth as artefacts of human behavior: intentional mutilation and accidental modification. In Kelly, M. A. and Larsen, C. S. (eds), *Advances in Dental Anthropology.* 357–78. New York.

Mogren, M. 1984. *Spetälska och spetälskehospital i Norden under medeltiden.* Lund.

Mortágua, A. 2005–2006. *Vikingarna från Slite. En osteologisk analys av ett vikingatida gravfält vid Slite torg, Gotland.* Stockholm.

Mortágua, A. 2006. Mutilated teeth. An analysis of eleven Vikings from Slite square, Gotland. unpublished MA dissertation, Stockholm University.

Mower, J. P. 1999. Deliberate ante-mortem dental modification and its implications in archaeology, ethnography and anthropology. *Papers from the Institute of Archaeology* 10, 37–53.

Mundal, E. 1989. Barneutbering. In Gunneng, H., Losman, B. and Møller Knudsen, B. (eds), *Kvinnors rosengård, Medeltidskvinnors liv och hälsa, lust och barnafödande. Föredrag från nordiska tvärvetenskapliga symposiet i Århus aug. 1985 och Visby sept. 1987.* Skriftserie från centrum för kvinnoforskning vid Stockholms universitet, 1, 122–34. Stockholm.

Nagmér, R. B. 1983. *Fjälkinge: boplatslämningar och gravar från sten- och järnålder: fornlämning 45, Fjälkinge socken, Skåne: arkeologisk undersökning 1980*, Stockholm.

Naidoo, J. and Willis, J. 2009. *Foundations for Health Promotion Public Health and Health Promotion Practice* (3rd edn). London.

Naumann, E., Krzewińska, M., Götherström, A. and Eriksson, G. 2014a. Slaves as burial gifts in Viking Age Norway? Evidence from stable isotope and ancient DNA analyses. *Journal of Archaeological Science* 41, 533–40. doi: 10.1016/j.jas.2013.08.022.

Naumann, E., Price, T. D. and Richards, M. 2014b. Changes in dietary practices and social organization during the pivotal late iron age period in Norway (AD 550–1030): Isotope analyses of Merovingian and Viking Age human remains. *American Journal of Physical Anthropology.* 155(3), 322–31. doi: 10.1002/ajpa.22551.

Nelsson, W. E. 1987. *Nelson Textbook of Pediatrics* (13th edn). Philadelphia.

Nevéus, C. 1974. *Trälarna i landskapslagarnas samhälle, Danmark och Sverige.* Studia historica Upsaliensia 58. Uppsala.

Nilsson, B. 2012. Vem döpte, när och var? In Borgehammar, S. (ed.), *Locus Celebris, Dalby kyrka, kloster och gård*, 129–50. Göteborg.

Östergren, M. 1983. Silverskatternas fyndplatser – farmennens gårdar. In Jansson, I. (ed.), *Gutar och Vikingar*, 34–48, Stockholm.

Östergren, M. 2003. Silverskatten från Hägvalds i Gerum. *Gotlands museums årsbok för 2003*, 104.

Paananen-Kemi, M. 2011. *Skärvor från en Husgrund. En fallstudie om Västergarnskeramiken från 2010 års fältundersökning*. Digitala Vetenskapliga Arkivet, DIVA: Unpublished.

Peets, J., Allmäe, R. and Maldre, L. 2010. Archaeological Investigations of pre-Viking Age burial boat in Salme village at Saaremaa. *Archeological Fieldwork in Estonia* 2010, 29–48.

Pentikäinen, J. 1968. *The Nordic Dead Child Tradition. Nordic Dead/Child Beings, a Study in Comparative Religion*. Helsingfors.

Pentikäinen, J. 1990. Child abandonment as an indicator of Christianization in Nordic countries, In Ahlbäck, T. (ed.), *Old Norse Finninsh Religious and Cultic Placenames*, 72–91. Åbo.

Peschel, E. 2014. Foreigners in Fröjel? A study of mobility on a Viking Age port of trade in Gotland, Sweden. Unpublished MSc School of Medicine, Boston University.

Petré, B. 1980. Björnfällen i begravningsritualen – statusobjekt speglande regional skinnhandel? *Fornvännen* 75, 5–14.

Petré, B. and Wigardt, M. 1973. *Björnklor i svenska gravfynd från järnåldern. Trebetygsuppsats vid Stockholms universitet*. Stencil.

Petersson, M. 2006. *Triberga kvarnbacke: ett gravfält från yngre järnålder: RAÄ 73, Hulterstad socken, Öland*. Kalmar.

Pettersson, H. 1966. Undersökning en av gravfältetet vid kopparsvik Visby. Premliminär redogörelse *Gotländskt arkiv*. S. 7–18. Visby.

Price, N. 2002. *The Viking Way: religion and war in late Iron Age Scandinavia*. Uppsala.

Price, T. D. 1985. Traces of late archaic substance in the Midwestern United States. *Journal of Human Evolution* 14, 449–60.

Price, T. D. 2013. Human mobility at Uppåkra. A preliminary report on isotopic proveniencing. In Hårdh, B. and Larsson, L. (eds), In *Folk, fä och fynd*, 163–75. Acta Archaeologica Lundensia 64. Lund.

Price, T. D., Arcini, C., Gustin, I., Drenzel, L. and Kalmring, S. 2018. Isotopes and human burials at Viking Age Birka and the Mälaren region, east central Sweden. *Journal of Anthropological Archaeology* 49, 19–38. https://doi.org/10.1016/j.jaa.2017.10.002.

Price, T. D., Frei, K. M., Dobat, A. S., Lynnerup, N. and Bennike, P. 2010. Who was in Harold Bluetooth's army? Strontium isotope investigation of the cemetery at the Viking Age fortress at Trelleborg, Denmark. *Antiquity* 85, 476–89.

Price, T. D., Naum, M., Bennike, P., Lynnerup, N., Frei, K. M., Wagnkilde, H. and Nielsen, F. O. 2013. Investigation of human provenience at the early medieval cemetery of Ndr. Grødbygård, Bornholm, Denmark. *Danish Journal of Archaeology* 1, 93–112.

Price, T. D., Peets, J., Allmäe, R., Maldre, L. and Oras, E. 2016. Isotopic provenancing of the Salme ship burials in Pre-Viking Age Estonia. *Antiquity* 90(352), 1022–37.

Price, T. D., Prangsgaard, K., Kanstrup, M., Bennike, P., Frei, K. M. 2015. Galgedil: isotopic studies of a Viking cemetery on the Danish island of Funen, AD 800–1050. *Danish Journal of Archaeology*, 3(2), 129–44. dx.doi.org/10.1080/21662282.2015.1056634.

Rastorfer, J-M. 2004. *On the Development of Kayan and Kayah National Identity: a study and an updated bibliography*. Bankok.

Reichart, P.A., Creutz, U. and Scheifele, C. 2008. Dental mutilations and associated alveolar bone pathology in African skulls of the anthropological skull collection, Charité, Berlin. *Journal of Oral Pathology & Medicine* 37(1), 50–5.

Resnick, D. and Niwayama, G. 1989. *Bone and Joint Imaging*. Philadelphia.

Robbins, G., Tripathy, V. M., Misra, V. N., Mohanty, R. K., Shinde, V. S., Gray, K. M. and Schug, M. D. 2009. Ancient skeletal evidence for leprosy in India (2000 B.C.) *PLos One* 4(5) https://doi.org/10.1371/journal.pone.0005669.

Roesdahl, E. 1988. *The Vikings*. London.

Romero, J. 1958. *Mutilaciones dentarias prehispánicas en México y América en general*. Serie Investigaciones 3. Mexico City.

Romero, J. 1970. Dental mutilation, trephination and cranial deformation. In Stewart, T. (ed.), *The Handbook of Middle American Indians* 9, 50–67. Austin.

Roslund, M. 2001. *Gäster i huset, Kulturell överföring mellan slaver och skandinaver 900–1300*. Lund.

Roslund, M. 2006 Kulturmötets konsekvenser. "Slaviseringen" av den skandinaviska keramiktraditionen. In Burström, I. M. (ed.), *Arkeologi och mångkultur*, 59–76. Södertörns Archaeological Studies 4.

Roslund, M. In press. n.d. Tacit knowing of thralls – style negotiation among the unfree in 11th and 12th C. Sweden, In Marcus, B. and Clack, T. (eds), In *Archaeologies of Contact and Hybridity*. Oxford, Oxford University Press.

Rubin, A. 1988. *Marks of Civilization: Artistic Transformations of the Human Body*. Los Angeles.

Samadelli, M., Miccoli, M., Melis, M. and Zink, A. R. 2015. Complete mapping of the tattoos of the 5300-year-old Tyrolean Iceman. *Journal of Cultural Heritage*. dx.doi.org/10.1016/j.culher.2014.12.005.

Saul, J. M. and Frank P. S. 1997. The Preclassic skeletons from Cuello. In Whittington, S. L. and Reed, D. M. (eds), *Bones of the Maya: Studies of Ancient Skeletons*, 181–95. Washington, DC.

Saul, F. P. and Hammond, N. 1973. A Classic Maya tooth cache from Lubaatun, British Honduras. In Graham, J. (ed), *Studies in Ancient Mesoamerica*, 31–5. Contribution of the University of California Archaeological Research Facility 18, Berkeley.

Saul, F. P. and Saul, J. M. 1991. The Preclassic population of Cuello. In Hammond, N. (ed), *An Early Maya Community in Belize*, 134–58. Cambridge.

Sawyer, P. 1999. *The Vikings*. Oxford.

Schroeder, H. and Haviser, J.B. 2012. The Zoutsteeg Three: Three new cases of African types of dental modification from Saint Martin, Dutch Caribbean. *International Journal of Osteoarchaeology* 24(6), 668–93. DOI: 10.1002/oa.2253.

Schulze, H. 1986. *Ett barngravfält från romersk järnålder: fornlämning 89 Bjärby, Kastlösa socken, Öland: arkeologisk undersökning 1972, 1975/76. Rapport – Riksantikvarieämbetet och Statens historiska museer. Undersökningsverksamheten*. Stockholm.

Scott, G.R. and Julie, R.B. 2008. Tooth-tool, use and yarn production in Norse Greenland. *Alaska Journal of Anthropology* 6(1–2), 253–64.

Segerstedt, T. 1907. *Mynts användning i dödskulten*. Lunds universitets årsskrift. Lund.

Sellevold, B. J., Hansen, U. L. and Baslev Jørgensen, J. 1984. *Iron Age Man in Denmark*. Prehistoric Man in Denmark 3. Copenhagen.

Sillen, A. and Kavanagh, M. 1982. Strontium and paleodietary research: a review. *Yearbook of Physical Anthropology* 25, 67–90.

Sjögren, K-G., Price, T. D. and Ahlström, T. 2009. Megaliths and mobility in south-western Sweden. Investigating relationships between a local society and its neighbours using strontium isotopes. *Journal of Anthropological Archaeology* 28, 85–101.

Sjøvold, T. 1990. Estimation of stature from long bones utilizing the line of organic correlation. *Human Evolution* 5(5), 431–47. https://link.springer.com/article/10.1007/BF02435593

Skre, D. 1998. *Herredømmet: bosetning og besittelse på Romerike 200–1350 e.Kr.* Oslo.

Smith, A. L. 1972. Dental decoration. In Smith, A. L. (ed.), *Excavations at Altar de Sacrificios: architecture, settlement, burials and caches.* 222–9. Papers of the Peabody Museum of Archaeology and Ethnology 62(2). Cambridge, MA.

Smith, H. 1996. *Maya Dental Mutilation. Dig It: Articles on Archaeology and the Maya on Ambergris Caye, Belize.* Belize.

Sörling, E. 1939. Penningväskor från vikingatid. *Fornvännen* 34, 45–57.

Sörling, E. 1945. Ännu en penningväska från Gotlands vikingatid. *Gotländskt arkiv:* 27–30.

Spindler, K. 2000. *Der Mann im Eis: neue sensationelle Erkenntnisse über die Mumie aus den Ötztaler Alpen.* Munich.

Steckel, R. 1995. Stature and standard of living. *Journal of Economic Literature* 33, 1903–40.

Steele, V. 2001. *The Corset: a cultural history.* New Haven.

Stenberger, M. 1964. *Det forntida Sverige.* Uppsala.

Stewart, T. D. and Tittererington, P. F. 1944. Filed Indian teeth from Illinois. *Journal of the Washington Academy of Science* 34, 317–21.

Stewart, T. D. and Tittererington, P. F. 1946. More filed Indian teeth from the United States. *Journal of the Washington Academy of Science* 36, 259–61.

Stolpe, H. 1872. *Naturhistoriska och archæologiska undersökningar på Björkö i Mälaren.* Öfversikt af Kongl. Vetenskaps-Akademiens Förhandlingar I. Stockholm.

Stolpe, H. 1876. Grafundersökningar på Björkö. *Tidskrift för antropologi och kulturhistoria* 1(10), 1–21.

Stolpe, H. 1878. Meddelanden från Björkö 1. En kristen begrafningsplats. *Månadsblad,* 671–84. Antikvarisk tidskrift för Sverige. Stockholm.

Stolpe, H. 1879a. *Årsrapport öfver Björkögrävningarna inlemnad den 8 jan. 1879.* Antikvarisk tidskrift för Sverige. Stockholm.

Stolpe, H. 1879b. *Årsrapport öfver Björkögrävningarna inlemnad den 15 dec. 1879.* Antikvarisk tidskrift för Sverige. Stockholm.

Stolpe, H. 1882. Grafundersökningar på Björkö i Mälaren år 1881. *Svenska Fornminnesföreningens Tidskrift* 5, 53–63.

Stolpe, H. 1989. Ett och annat om Björkö i Mälaren. *Ny illustrerad tidning,* 461–4.

Strömberg, M. 1950. Skelettgravar vid Fjälkinge nr 3564, Fjälkinge socken. Riksantikvarieämbetet. Stockholm. Unpublished.

Svanberg, F. 2003. *Decolonizing the Viking Age.* Acta Archaeologica Lundensia. Lund.

Svanberg, F. 2016. Who were the Vikings? In Andersson, G. (ed.), *We Call Them Vikings,* 16–22. Stockholm.

Syse, B. 2003. *Långfredagsslaget: en arkeologisk historia.* Uppsala.

Tanner, J. M., Hayashi, T., Preece, M. A. and Cameron, N. 1982. Increase in length of leg relative to trunk in Japanese children and adults from 1957 to 1997: comparison with British and with Japanese Americans. *Annals Human Biology* 9, 411–23.

Tiesler, V. 1999. Head shaping and dental decoration among the Ancient Maya: archeological and cultural aspects. *Unpublished paper presented at the 64th Meeting of the Society of American Archaeology.* Chicago, IL.

Tiesler, V., Ramírez, M. and Oliva, I. 2005. Técnicas de decoración dental en México. *Actuealidades Arqueológicas* 2, 18–24.

Tiesler Blos, V. 2001. *Decorationes dentales entre los antigues Mayas.* Mexico City.

Thunmark-Nylén, L. 1995. *Die Wikingerzeit Gotlands I.* Vitterhets historie. och antikvitets akademien. Stockholm

Thunmark-Nylén, L. 2000a. *Die Wikingerzeit Gotlands IV:1.* Vitterhets historie. och antikvitets akademien Stockholm

Thunmark-Nylén, L. 2000b. *Die Wikingerzeit Gotlands IV:2* Vitterhets historie. och antikvitets akademien. Stockholm

Thunmark-Nylén, L. 2004. Visby – ett pussel med gamla och nya pusselbitar. *Fornvännen* 4, 285–97.

Thunmark-Nylén, L. 2006. *Die Wikingerzeit Gotlands* 3(2), Text Kungl. Vitterhets historie och antikvitets akademien. Stockholm.

Toplak, M. 2016. *Das wikingerzeitliche Gräberfeld von Kopparsvik auf Gotland: Studien zu neuen Konzepten sozialer Identitäten am Übergang zum christlichen Mittelalter.* Tübingen.

Tracy, M. R., Dormans, J. P. and Kusumi, K. 2004. Klippel-Feil syndrome: clinical features and current understanding of etiology. *Clinical Orthopaedics and Related Research* 424, 183–90.

Übelaker, D. H. 1987. Dental alteration in prehistoric Ecuador. A new example from Jama-coaque. *Journal of Washington Academy of Sciences* 77(2), 76–80.

Van Reenen, J. F. 1986. Tooth mutilating and extraction practices amongst the peoples of south west Africa (Namibia). In Singer, R. and Lundy, J. K. (eds), *Variation, Culture and Evolution in African Populations: Papers in Honour of Dr. Hertha de Villiers,* 159–69, Johannesburg.

Vedeler, M. 2014. *Silk for the Vikings.* Ancient Textiles Series 15. Oxford.

von Jhering, H. 1882. Die kunstliche Deformirung der Zähne. *Zeitgeshrift für Etnologie* 14, 213–62.

Vos, A. 2005. En osteologisk sammanställning av 141 individer från den vikingatida hamnplatsen vid Fröjel på Gotland. Osteologisk analys av 2004 års gravar från det vikingatida gravfältet på Bottarve 1:17 i Fröjel. Unpublished MA thesis. Stockholm University.

Vukovic, A., Bajsman, A., Zukic, S. and Secic, S. 2009. Cosmetic dentistry in ancient times – a short review. *Bulletin of the International Association for Paleodontology* 3, 9–13.

Welinder, S., Pedersen, E. A. and Widgren, M. 1998. *Det svenska jordbrukets historia. Bd 1. Jordbrukets första femtusen år: 4000 f.Kr.- 1000 e.Kr.* Stockholm.

Westholm, G. 1988. Visby – Böndernas hamn och handelsplats – Visbysamhällets uppkomst och utbredning under förhistorisk tid och äldre medeltid. *Medeltidsstaden* 72(2).

WHO 1946. Constitutions of WHO: principals. WHO http://www.who.int/about/mission/en/

Wikander, S. 1978. *Araber, vikingar och väringar.* Lund.

Wilhelmson, H. and Ahlström, T. 2015. Iron Age migration on the island of Öland: apportionment of strontium by means of Bayesian mixing analysis. *Journal of Archaeological Science* 64: 30–45. DOI: 10.1016/j.jas.2015.09.007.

Wilhelmson, H. and Price, D. T. 2017. Migration and integration on the baltic island of Öland in the Iron Age. *Journal of Archaeological Science:* Reports 12, 183–96.

Willey, G. R. William, R. B. Jr, Glass, J. B. and Gifford, J. C. 1965. *Prehistoric Maya Settlement in Belize Valley.* Harvard University Paper 54. Cambridge, MA.

Williams, J. S. and White, C. D. 2006. Dental modification in the Postclassic population from Lamanai, Belize. *Ancient Mesoamerica* 17: 139–51.

Yrwing, H. 1992. Visbysamhällets uppkomst – ett inlägg mot Gun Westholms framställning i Medeltidsstaden. I. *Fornvännen* 87: 191–200.

Zachrisson, T. 1994. The Odal and its manifestation in the landscape. *Current Swedish Archaeology* 2: 219–38.

Zachrisson, T. 2014. Trälar fanns – att synliggöra ofria 550–1220 e.Kr. i Sverige. In Carlie, A (ed.), *Att befolka det förflutna. Fem artiklar om hur vi kan synliggöra människan och hennes handlingar i arkeologiskt material*, 72–91. Stockholm.

Zumbroich, T. J. 2011. To strengthen the teeth and harden the gums. Teeth blackening as medical practice in Asia, Micronesia and Melanesia. *Ethnobotany Research and Applications* 9: 97–113.

Zumbroich, T. J. and Salvador-Amores. A. 2009. When black teeth were beautiful – the history and ethnography of dental modifications in Luzon, Philippines. *Studia Asiatica* 10: 125–65.

Zumbroich, T. J. and Salvador-Amores, A. 2010. Gold work, filing and blackened teeth:dental modifications in Luzon. *Cordillera Review* 2(2): 3–42.

Websites

http://www.birkavikingastaden.se/2016/04/birkaportalen/
http://www.socialstyrelsen.se/ovanligadiagnoser/medfoddspondyloepifysealdyspla